T0135771

Efficient reheating concepts for a reverse-flow methane reformer

Von der Fakultät Energie-, Verfahrens- und Biotechnik
der Universität Stuttgart zur Erlangung der Würde
eines Doktors der Ingenieurwissenschaften (Dr.-Ing.)
genehmigte Abhandlung

vorgelegt von
Carlos Tellaeche Herranz
geboren in Santurtzi, Spanien

Hauptberichter : Prof. Dr.-Ing. U. Nieken
Mitberichter : Prof. Dr.ir. M. van Sint Annaland
Tag der mündlichen Prüfung:
20.09.2013

Institut für Chemische Verfahrenstechnik
der Universität Stuttgart
2013

Bibliografische Information der Deutschen Nationalbibliothek

Die Deutsche Nationalbibliothek verzeichnet diese Publikation in der Deutschen Nationalbibliografie; detaillierte bibliografische Daten sind im Internet über http://dnb.d-nb.de abrufbar.

D 93

ISBN 978-3-8325-3641-1

Logos Verlag Berlin GmbH
Comeniushof, Gubener Str. 47,
10243 Berlin
Tel.: +49 (0)30 42 85 10 90
Fax: +49 (0)30 42 85 10 92
INTERNET: http://www.logos-verlag.de

Vorwort

Die vorliegende Arbeit entstand im Rahmen meiner Tätigkeit als Wissenschaftlicher Mitarbeiter am Institut für Chemische Verfahrenstechnik der Universität Stuttgart.

Herrn Prof. Dr. Nieken, der mir die Arbeit am Institut erst ermöglicht hat, danke ich für seine förderlichen Ratschläge und Diskussionen sowie das von ihm geschenkte Vertrauen. Die Kultur des Instituts und der wertvolle Freiraum haben mich nachhaltig geprägt.

Herr Prof. Dr. M. van Sint Annaland danke ich für das Interesse an meiner Arbeit und für die freundliche Übernahme des Mitberichts.

Herrn Prof. Dr. Eigenberger für die offenen und konstruktiven Diskussionen, die mich und meine Arbeit immer bereichert haben. Ebenso dafür bedanke ich mich bei Herrn Dr. Grigorios Kolios, den ich als wertvollen Ratgeber und Freund außerordentlich schätze.

Bernd Glöckler für seine langjährige Unterstützung, die er mir bereits während meiner Studentenzeit am Institut gewährt hat. Sowohl seine Begeisterung und Faszination für unseren Beruf als auch seine Arbeit, die als Grundlage zu dieser Dissertation dient, sind mir ein Vorbild.

Allen Institutskolleginnen und Kollegen danke ich für die freundliche und fruchtbare Atmosphäre, die aus der Arbeit am ICVT eine unersetzbare Erfahrung gemacht haben. Ganz besonders Franz Keller, Andreas Freund, Christian Spengler, Manuel Huber, Philipp Günther, Stefan Dwenger, Jens Bernnat und Karin Hauff, die mir über unser Arbeitsverhältnis hinaus ihre Freundschaft geschenkt haben.

Ich danke allen Institutsmitarbeitern, die im Hintergrund dafür sorgen, dass sich alle Zahnräder unaufhörlich drehen und damit eine Arbeit wie die Vorliegende erst ermöglichen. Herrn Dr. Sorescu für seine Unterstützung während und nach meiner Zeit am Institut, Herrn Lorenz und den Werkstattmitarbeitern für den unabdingbaren Beitrag zum Erfolg der experimentellen Arbeiten, Holger Aschenbrenner für seine immer freundlichen und geduldigen Hilfestellungen, Robert Fettig für die langjährige hervorragende Zusammenarbeit sowie Katrin Hungerbühler für ihre Fürsorge.

Darüber hinaus danke ich allen Studenten, die mit Engagement einen Beitrag für meine Arbeit geleistet haben. Stellvertretend für alle hebe ich Falk Eichhorn, Markus Schmudde, Andreas Dieter, Michael Speidel und Pia Patzelt hervor.

An meine Eltern richte ich mein endloses, aufrichtiges und liebstes Dankeschön. Sie haben die Fundamente gesetzt und zusammen mit meinen Schwestern und Dilara mir die Kraft und Motivation geschenkt, die es möglich gemacht haben, dass ich diese Widmung heute schreiben darf.

Heidelberg, September 2013

a mi familia

Contents

Notation

ATR	Autothermal reforming
BDF	Backward Differentiation Formula
CC	Combustion chamber
CFD	Computational Fluid Dynamics
CPOM	Catalytic partial oxidation of methane
CRR	Combustion and reforming reactions
CSTR	Continuously stirred tank reactor
DAE	Differential algebraic equation
DIANA	Dynamic simulation and nonlinear analysis
DPO	Direct partial oxidation mechanism
EDM	Eddy dissipation model
FLOX	Flameless oxidation
GHR	Gas heated reformer
HFR	High flux steam reformer
HTCR	Haldor Topsøe convection reformer
HTER	Haldor Topsøe exchange reformer
ICVT	Institut für Chemische Verfahrenstechnik
LDF	Linear Driving Force
MS	Main stream
MSR	Methane steam reforming
NDIR	Non dispersive infrared
NG	Natural gas
NLE	Nonlinear equation
ODE	Ordinary differential equation
PAC	Programmable Automation Controller
PDE	Partial differential equation
PSA	Pressure swing adsorption
ProMoT	Process Modeling Tool
RF	Real factor

RFR	Reverse flow reactor	
SF	Side feed	
WGS	Water-gas shift	

Latin Letters

A	Reactor cross section	m^2
$a^v_{j,k}$	Specific surface between phases j and k	m^2/m^3
c_k	Specific heat of solid phase k	$kJ/kg \cdot K$
$c_{p,k}$	Heat capacity of phase k	$kJ/kg \cdot K$
D	Diameter	m
D_{eff}	Effective diffusivity	$kg/m \cdot s$
d_h	Hydraulic diameter	m
d_p	Particle diameter	m
$\Delta h_{C,i}$	Molar combustion enthalpy of component i	kJ/mol
$\Delta h_{r,i}$	Molar reaction enthalpy of reaction i	kJ/mol
$k^{j,k}$	Overall heat transfer coefficient between phases j and k	$kW/m^2 \cdot K$
K_V	Recirculation rate	
L	Length of the reactor packing	m
m	Total mass	kg
$\dot{M}$	Mass flow	kg/s
$\dot{m}$	Mass flow density	$kg/m^2 \cdot s$
MW_j	Molecular weight of component j	$kg/kmol$
N_j	Mole flow of component j	$kmol/s$
$\dot{N}_j$	Mole flow density of component j	$kmol/m^2 \cdot s$
Nu	Nusselt number	-
Pe	Peclet number	-
Pr	Prandtl number	-
$\dot{q}$	Heat flux density	kW/m^3
Q	Heat flux	kW
Re	Reynolds number	-
Re_p	Particle Reynolds number	-
r_i^{cat}	Reaction rate	$mol/m^3 \cdot s$

t	Time	s
T_k	Temperature of phase k	K
ΔT_{ad}	Adiabatic temperature increase/drop	K
t_{mix}	Mixing time	s
v	Linear velocity	m/s
V_k	Volume of phase k	m^3
w_j	Mass fraction of component j	kg_j/kg_{tot}
w	Front propagation velocity	m/s
X_j	Conversion of component j	–
z	Length, spatial coordinate	m
z_{shift}	Specific displacement of a temperature profile	m

Greek Letters

α_k	Convective heat transfer coefficient of phase k	$kW/m^2 \cdot K$
ε	Void fraction in the packing	m^3_{void}/m^3_{tot}
η_{therm}	Energetic efficiency	%
λ	Air-fuel equivalence ratio	-
λ_k	Thermal conductivity of phase k	$kW/m \cdot K$
λ_{eff}	Effective thermal conductivity	$kW/m \cdot K$
μ_k	Dynamic viscosity of phase k	$Pa \cdot s$
$\nu_{i,j}$	Stoichiometric coefficient of component j in reaction i	-
ϕ	Fuel/air equivalence ratio	-
Φ_k	Phase fraction of phase k	m^3_k/m^3_{tot}
ρ_k	Density of phase k	kg/m^3
σ	Standard deviation	-
τ	Residence time	s
$\tau_{semicycle}$	Semicycle duration in periodic operation. *Semicycle* stays for production (*prod*) or regeneration (*reg*)	s
Θ	Corrected equivalence ratio	-
ξ_i	Extent of reaction	mol_i

Superscripts

$+$	Inlet flow
endo	Endothermic

Subscripts

active	Catalytic active packing
air	Air flow
amb	Surrounding, ambient
cat	Packed-bed
cc	Combustion chamber
comb	Combustion reaction
crit	Critical
effluent	Reactor effluent
eff	Effective
endo	Endothermic
end	End of a process
eq,equil	Equilibrium conditions
exo	Exothermic
exp	Experimentally measured
fuel	Fuel flow
gas	Gas bulk
g	Gaseous phase
inert	Inert flow / packing
inlet	Inlet boundary of the integration domain
ins	Insulation
in	Mass/Energy flow entering a control volume
i	Reaction index
jet	Injected flow through one single nozzle
j	Chemical species index
k	Reactor phase
low	Low, lowest

max	Maximal value
mix	Mixing, mixture
MS	Main stream
outlet	Outlet boundary of the integration domain
out	Mass/Energy flow leaving a control volume
product	Product flow leaving a control volume
prod	Production semicycle
rea	Reaction zone
rec	Recirculated
ref	Reforming reaction or semicycle
reg	Regeneration semicycle
rw	Reactor wall
R	(Exothermic) Reaction front
shock	Shock region in the endothermic reaction front
SS	Side stream
stat	Stationary
s	Solid phase
theor	Theoretical
therm	Thermal process
tot	Total, cumulative
up	Upper, highest
LHV	Lowest heating value

Zusammenfassung

Die vorliegende Arbeit widmet sich der Weiterentwicklung des von B. Glöckler untersuchten Strömungsumkehrreaktorkonzeptes für die stark endotherme Dampfreformierung von Methan [1, 2]. Beim Strömungsumkehrreaktor (RFR, aus dem Englischen *reverse-flow reactor*) handelt es sich um einen wärmeintegrierten Reaktor. Dieser zeichnet sich dadurch aus, dass die Wärmekapazität des katalytischen Festbettes als Energiereservoir eingesetzt wird. Dieser Reaktortyp wird periodisch durch die Umkehr der Strömungsrichtung in definierten Zyklen betrieben, was eine regenerative Wärmerückgewinnung ermöglicht. Dafür wird die während eines Zyklus in der Austrittsrandzone des Festbettes gespeicherte Energie im Laufe des nachfolgenden Zyklus auf den Zulauf übertragen.

Die Betriebsweise des Strömungsumkehrreaktors wird bereits für die autotherme Durchführung leicht exothermer Reaktionen angewandt, beispielsweise für die katalytische Aufbereitung industrieller Abluft. Darüber hinaus stellt der hervorragende Energieaustausch zwischen dem Gas und der Schüttung ein großes Potenzial dar, das für die energetische Koppelung stark endothermer und exothermer Reaktionen genutzt werden kann. Dies ist beim von Glöckler untersuchten Strömungsumkehrreaktorkonzept der Fall, bei dem die Koppelung zwischen der endothermen Methan-Dampfreformierung und der Verbrennung eines Brenngases in einem regenerativen, asymmetrischen Betriebsmodus erfolgt. Die Zulaufzusammensetzung wird für die Durchführung der endothermen (Produktion) und der exothermen Reaktionen (Regeneration) mit jeder Strömungsumkehr geändert.

Bei diesem Konzept wird der Reaktor angefahren, indem das adiabate Festbett auf Temperaturen um die 1000 °C vorgeheizt wird, die für das Erreichen hoher Methanumsätze während des Produktionsschrittes erforderlich sind. Die Methan-Dampfreformierung läuft während des Produktionsschritts ab, wobei sich eine selbst schärfende, wandernde Reaktions- und Temperaturfront ausbildet. Diese Reaktionsfront verbraucht für die Durchführung der endothermen Reformierung die in der Katalysatorschüttung gespeicherte Wärme. Solange die Front diese Schüttung nicht verlässt, wird bei der Reformierungsreaktion Vollumsatz erreicht. Gelangt die Front an das Ende der katalytischen Schüttung, wird die Produktionsphase beendet, woraufhin das Festbett während der sogenannten Regenerationsphase wiederaufgeheizt wird.

Um die Schüttung auf die Ausgangstemperatur wieder aufzuheizen, muss die Strömungsrichtung umgekehrt werden. Sind die Wärmekapazitätsströme in der Produktions- und der Regenerationsphase idealerweise identisch, so ist eine vollständige regenerative Wärmeintegration beider Phasen möglich. Die durch die Reformierung verbrauchte Energie kann durch eine Verbrennung der Katalysatorschüttung zugeführt werden. Wie Glöckler in [1] zeigte, muss die Wärme an mehreren über der Schüttung verteilten Stellen zugeführt werden, um das Festbett auf eine einheitliche Ausgangstemperatur wieder aufzuheizen.

In dem von Glöckler untersuchten Konzept wird der Reaktor mit einem brennstoffhaltigen Regenerationsgas durchströmt. Gleichzeitig wird Luft an mehreren inerten Abschnitten entlang des Festbettes zugeführt. Das brennstoffhaltige Gas und die Luft sollen sich in den inerten Schüttungsabschnitten vermischen und in der darauffolgenden katalytischen Schüttung miteinander reagieren, sodass die Teilverbrennung des Brenngases die für die Aufheizung der Schüttung notwendige Energie liefert.

Die direkte Wärmefreisetzung zur Aufheizung der Schüttung bildet gegenüber den konventionellen Reformierungstechnologien einen wesentlichen Vorteil, welche aus Rohrbündelapparaten, die im Strahlungsofen von außen beheizt werden, bestehen. Darüber hinaus ermöglicht die oben genannte regenerative Wärmeintegration das Erreichen einer höheren thermischen Effizienz und zeichnet das Konzept für die dezentrale Wasserstoffherstellung aus.

Eine experimentelle Validierung des Konzeptes wurde von Glöckler et al. in [2] veröffentlicht. Diese Ergebnisse stellen den Ausgangspunkt dieser Arbeit dar und beschreiben das Auftreten von lokalen Temperaturspitzen an den Lufteinspeisungsstellen durch spontane Verbrennung. Diese hohen Temperaturen führen zur Schädigung der Lufteinspeisungen, der Schüttung sowie zu einem unregelmäßigen Temperaturprofil am Ende der Regenerationsphase. Aus diesem Grund verfolgt die Arbeit das Ziel, eine alternative Regenerationsstrategie, die die genannten Probleme vermeidet, auszuarbeiten.

Der Lösungsansatz dieser Aufgabe bildet die in industriellen Brennern bereits etablierte flammlose Oxidationstechnologie (FLOX®). Diese Brenner werden mit Luftüberschuss betrieben und sind so konzipiert, dass die Verbrennungsluft, das Brenngas und die Verbrennungsabgase vollständig miteinander vermischt werden. Bei ausreichend hohen Brennkammertemperaturen hat dies zufolge, dass die Verbrennung homogen und ohne sichtbare Flammenbildung verläuft. Ausschlaggebend für eine flammlose Verbrennung ohne Auftreten von Übertemperaturen ist eine ausreichende Rückvermischung der Verbrennungsabgase mit der Luft und dem Brennstoff, die frisch zugeführt werden. Diese Rückvermischung wird in der Literatur mit einem Rezirkulationsverhältnis von Abgas und Frischgas K_V charakterisiert, welches Werte größer als 2 aufweisen soll. Um dieses Prinzip auf das Strömungsumkehrreaktorkonzept zu übertragen, müssen schüttungslose Brennkammern entlang des

Reaktors eingebettet werden. Dort soll während der Regenerationsphase die oben beschriebene flammlose Verbrennung stattfinden, sodass eine intensive Vermischung der Verbrennungsluft, des Brennstoffes und des Verbrennungsabgases erforderlich ist. Für die Konzipierung und Entwicklung einer solchen Brennkammer wurde ein breites Spektrum an Untersuchungsmethoden herangezogen: Strömungs-, Vermischungs- und Reaktionsverhalten der Brennkammern wurden anhand von CFD-Simulationen und Strömungssichtbarmachungs- und Verbrennungsexperimenten an Modellbrennkammern untersucht.

Ausgangspunkt für die Entwicklung der Brennkammer ist das bisherige Konzept, bei dem der Brennstoff verdünnt in einem Trägergas zugeführt wird. Die Verbrennungsluft wird in unterstöchiometrischem Verhältnis zum Brennstoff in die Brennkammer so eingespeist ($\lambda < 1$), dass die erforderliche Rückvermischung des Brennkammerinhaltes gewährleistet wird. Dafür werden die Geometrie und die Anordnung der Luftdüsen sowie das günstige Verhältnis zwischen Luft- und Trägergasmenge untersucht. Dies erfolgt sowohl mittels Strömungssimulationen als auch optisch in einem Modell der Brennkammer, das basierend auf Reynolds-Ähnlichkeitsgesetzen mit Wasser betrieben wird. Anschließend werden Verbrennungsversuche in einem Quarzglasreaktor durchgeführt und die Temperaturverteilung in der Brennkammer bewertet. Bei der entwickelten Brennkammerkonfiguration wird die Verbrennungsluft durch eine im Ausgangsbereich der Kammer zentral angebrachte Luftdüse eingespeist. Dies erfolgt in entgegengesetzter Richtung zum Hauptstrom, der aus dem Brennstoff und dem Trägergas besteht.

Die auf Basis der unterschiedlichen Untersuchungsmethoden gewonnenen Ergebnisse liefern die einheitliche Aussage, dass sowohl das Design der Luftdüsen als auch ein hohes Verhältnis zwischen der Düsengasmenge und der Trägergasmenge entscheidend für eine flammlose Verbrennung sind. Kann dieses hohe Verhältnis im dynamischen Strömungsumkehrbetrieb nicht immer gehalten werden, ist die Flammenbildung am Austritt der Luftdüsen nicht auszuschließen. Die Entstehung von Flammen erfolgt allerdings in dem von der Brennkammer ausgebildeten und mit Gas gefüllten Leerraum, weshalb die Übertemperaturen kein Risiko für die Katalysatorschüttung und die Reaktormaterialien darstellen.

Im Strömungsumkehrbetrieb spielt das Zündverhalten der Brennkammer eine wichtige Rolle. Während der Produktionsphase werden das Festbett und die Brennkammern abgekühlt, so dass die Brennkammertemperatur am Anfang der Regenerationsphase unterhalb der Zündtemperatur der homogenen Verbrennung liegt. Die Verbrennungsluft und das Brenngas vermischen sich in der Brennkammer und zünden in der katalytischen Schüttung stromabwärts. Dort findet die Reaktion statt, bis die Temperatur in der Brennkammer die

Zündtemperatur der homogenen Reaktion erreicht. So wie der Vorgang beschrieben ist, findet ein Zündverzug statt. Allerdings entstehen dank der guten Vermischung von Brenngas und Luft vor Eintritt in die Katalysatorschüttung keine Übertemperaturen. Somit ist eine kontrollierte, gleichmäßige Verbrennung mit annähernd konstanter adiabater Temperaturerhöhung auch unter dynamischen Betriebsbedingungen möglich.

Der Betrieb der Kammer bei unterstöchiometrischen Verhältnissen weist jedoch eine starke Empfindlichkeit gegenüber dem eingesetzten Brennstoff auf. Wird ein kohlenstoffhaltiger Brennstoff verwendet, findet eine Partialoxidation statt. Während ein Teil des Brennstoffes am Eintritt des Katalysators verbrennt und einen Temperaturanstieg hervorruft, folgt daraufhin eine endotherme Reformierung des restlichen Brennstoffes und die katalytische Schüttung kühlt ab. Dieses Verhalten erklärt mitunter die ungleichmäßigen Temperaturprofile am Ende der Regenerationsphase bei Verwendung von methanhaltigen Brennstoffen, wie dies bei der von Glöckler oben angesprochenen experimentellen Validierung des Strömungsumkehrkonzepts der Fall war.

Um den Strömungsumkehrreaktor nicht nur energetisch, sondern auch stofflich in ein dezentrales Verfahren zu integrieren, soll das Abgas einer Druckwechseladsorption (DWA-Abgas oder PSA-Abgas aus dem Englischen *pressure swing adsorption*) für die Wasserstoffaufbereitung während der Regeneration als Brenngas eingesetzt werden. Da dieses Abgas signifikante Methananteile enthält, ergibt sich die Notwendigkeit eine alternative Regenerationsstrategie zu entwickeln, bei der die Methanreformierung effektiv unterdrückt wird.

Ein möglicher Ansatz könnte darin bestehen, den Reaktor mit Luft als Regeneriergas durchzuspülen und das Brenngas in die Brennkammern einzudüsen, sodass überstöchiometrische Bedingungen gehalten werden ($\lambda > 1$). Jedoch könnte sich die Oxidierung des Katalysators während der nachfolgenden Produktionsphase als negativ erweisen, weshalb die Entwicklung einer Regenerationsstrategie, bei der die Verbrennung in jeder Brennkammer unter stöchiometrischen Bedingungen ($\lambda = 1$) abläuft, verfolgt wird. Der Reaktor wird in diesem Konzept mit einem Trägergas durchströmt, während die Luft und das Brenngas im stöchiometrischen Verhältnis und voneinander getrennt in die Brennkammer zugeführt werden. Wie vorhin beschrieben, ist eine gute Vermischung der drei Ströme für eine flammlose Verbrennung entscheidend. Eine Untersuchung der dafür erforderlichen Betriebsparameter und des Verbrennungsverhaltens erfolgt experimentell in einer metallischen Brennkammer. Im Gegensatz zu der bisherigen Auslegung der Brennkammer wird bei diesem Konzept das Brenngas auf mittlerer Höhe der Kammer seitlich zugeführt.

Die Ergebnisse bestätigen die Vorteile, die mit einer stöchiometrischen Verbrennung einhergehen. Diese Vorteile beziehen sich zum einen auf die erfolgreiche Unterdrückung von unerwünschten Nebenreaktionen und zum anderen auf eine deutliche Vereinfachung des Regelungskonzeptes der Brennkammer für den Strömungsumkehrbetrieb. Auf Basis der experimentellen Daten, der Analyse der Temperaturhomogenität und dem Brennstoffumsatz kann ein Betriebsfenster definiert werden, bei dem hohe Umsätze ($> 90\%$) und geringe Temperaturdifferenzen in der Kammer erreicht werden. Dieses Betriebsfenster ist allerdings durch das Eindüsungskonzept des Brennstoffes stark begrenzt und wird für den Betrieb der Brennkammern im Strömungsumkehrreaktor berücksichtigt.

Die experimentellen Beobachtungen werden durch detaillierte Strömungssimulationen der Brennkammer bestätigt. Diese zeigen, dass die Verdünnung des Brenngases mit dem Trägergasstrom eine entscheidende Rolle beim Erreichen einer homogenen Verbrennung spielt. Demzufolge sind die Unterdrückung der Flammenbildung und eine homogene Temperaturverteilung in der Kammer mit einer vom Trägergasstrom getrennten Brenngasdosierung schlechter als im ursprünglich untersuchten Fall. Wie bereits beschrieben und experimentell bestätigt, tritt jedoch die Flammenbildung innerhalb der Brennkammer auf, weshalb weder der Katalysator noch die Reaktoreinbauten beschädigt werden können.

Die Validierung des Regenerationskonzeptes unter Strömungsumkehrbedingungen berücksichtigt die oben beschriebenen Erkenntnisse. Bei dem dafür konzipierten Stahl-Reaktor sind drei Brennkammern eingebaut. Der Reaktor wird bei Reformierlasten im Bereich zwischen 2 und $5\,kW_{LHV,H_2}$ betrieben. Für die Regeneration werden sowohl H_2 als auch synthetisches DWA-Abgas verwendet. Brenngas und Luft werden im stöchiometrischen Verhältnis in die Kammern geregelt zugeführt, sodass die Brennkammern auf einen konstanten Temperatursollwert aufgeheizt werden.

Die Betriebsparameter werden so gewählt, dass eine effektive Wärmerückgewinnung erzielt werden kann. Um die Dauer der Produktions- und Regenerationsphasen gleich zu halten, müssen die Gasmengen so gewählt werden, dass sich in beiden Phasen idealerweise gleiche Wärmekapazitätsströme ergeben. Gleichzeitig beeinflusst dies die erzielbare Wärmerückgewinnung, die an der Temperaturdifferenz zwischen Zulauf und Ablauf gemessen werden kann. Während die Zulauftemperaturen etwa 150 °C betragen, wird der Reaktor so betrieben, dass maximale Temperaturen von 300 °C im Ablauf nicht überschritten werden.

Bei den im Strömungsumkehrreaktor erzielten Ergebnissen wird ersichtlich, dass eine effiziente Wiederaufheizung des Festbettes und die erfolgreiche Vermeidung von schädigenden Temperaturspitzen mit der erarbeiteten Regenerationsstrategie möglich sind. Im Gegensatz zu der von Glöckler beschriebenen experimentellen Validierung des Reaktorkonzepts in [2] können dadurch am Ende der Regeneration gleichmäßige Temperaturprofile um die 900 °C erreicht werden. Selbst unter Betriebsbedingungen, bei denen Temperaturinhomogenitäten in den Brennkammern nicht unterdrückt werden können, gleichen sich die Temperaturen aufgrund der geringen Wärmekapazität der Brennkammern unmittelbar nach Beginn der Produktion aus.

Die Leistung des neuen Reaktorkonzepts erweist sich im Auslegungsbereich robust und gegen Lastwechsel unempfindlich. Bei mäßiger Last werden Gleichgewichtsumsätze entsprechend der Ablauftemperatur aus der katalytischen Schüttung von über 90% erreicht. Darüber hinaus zeigen die Temperaturprofile an den Reaktorrändern eine gute Wärmerückgewinnung. Allerdings betragen die erreichten thermischen Wirkungsgrade lediglich 40% bis 50%, wofür die unvermeidbaren Wärmeverluste der experimentellen Anlage und der später diskutierten Zunahme des Wärmekapazitätsstroms in Strömungsrichtung in beiden Phasen verantwortlich sind. Dieser letzte Effekt ist unter anderem von den physikalischen Eigenschaften des Gases und dem eingesetzten Brennstoff abhängig. Durch eine geeignete Anpassung der Betriebsparameter, d.h. der Produktions- und Regenerationsmassenströme, kann dieser Effekt teilweise kompensiert werden. Nach der erfolgreichen experimentellen Validierung des Konzeptes eignet sich für eine detaillierte Analyse des Systems und die Untersuchung von Variationsmöglichkeiten des Strömungsumkehrkonzepts der Einsatz von Simulationsrechnungen.

Für diesen Zweck wird ein mathematisches Modell entwickelt, das die Schüttung als Mehrphasen-Pfropfenströmungsmodell mit axialer Dispersion beschreibt. Die Brennkammern werden dagegen als ideal durchmischter Rührkessel unter Vollumsatzannahme in der Gasphase während der Regeneration modelliert. Die Simulation erfolgt mit der Software DIANA, die am Max-Planck-Institut für Dynamik komplexer technischer Systeme in Magdeburg entwickelt wurde.

Wie bereits genannt, setzt eine optimale regenerative Wärmerückgewinnung einen konstanten und gleichen Wärmekapazitätsstrom in beiden Strömungsrichtungen voraus. Allerdings erhöht sich die Wärmekapazität des Gases während der Produktion entlang des Reaktors infolge der Reformierungsreaktion um etwa 35%. Während der Regenerationsphase erhöht sich die Wärmekapazität durch die Luft- und Brenngaszugabe beim Einsatz vom DWA-Abgas um ca. 15% und beim Einsatz von Wasserstoff als Brennstoff um 30%. Demzufolge sind die austretenden Wärmekapazitätsströme deutlich größer als die eintretenden, weshalb nur eine eingeschränkte energetische Rückgewinnung möglich ist. Darüber hinaus ist die Wärmekapazität des Reformats etwa doppelt so hoch wie die des Regenerationsgases, weshalb ein Massenstromverhältnis von Regenerier- zu Reformiergasstrom von etwa 2 erforderlich ist. Die Simulationsergebnisse zeigen, dass der höchste thermische Wirkungsgrad bei gleichzeitigem Vollumsatz von Methan bei einem Massenstromverhältnis von etwa 2.5 erreicht werden kann. Obwohl anhand von Simulationsergebnissen Wirkungsgrade von über 80% erwartet werden, ist die Diskrepanz zwischen den Wärmekapazitätsströmen durch Anpassung der Massenstromverhältnisse oder der Periodendauer trotzdem nicht vollständig auszugleichen.

Ein weiterer Aspekt, der mittels Simulation untersucht werden kann, ist eine alternative Reaktorkonfiguration und Strukturierung des Festbettes. Der bisher untersuchte Strömungsumkehrreaktor besteht im reformiergasseitigen Eingangsbereich aus einem katalytischen Schüttungsabschnitt. Dagegen ist der Ausgangsbereich mit inerter Schüttung gefüllt, um eine Rückreaktion beim Abkühlen des Produktgases zu verhindern. Da aber das Reformiergas im Eingangsbereich aufgeheizt wird und die Temperatur der Schüttung in der Regel nicht hoch genug ist, um einen nennenswerten Beitrag zum Gesamtumsatz zu leisten, wird der Vorschlag von Glöckler et al. in [3] aufgegriffen und auch im Eingangsbereich eine inerte Vorschüttung vorgesehen. Reicht die während der Regenerationsphase in diesem Abschnitt gespeicherte Energie aus, um eine hohe Temperatur des Reformiergases vor seinem Eintritt in die Katalysatorschüttung zu gewährleisten, so entsteht an dieser Stelle eine stationäre Reaktionszone. In dieser findet Teilumsatz statt, was sich auf einen steilen Temperaturabfall auswirkt. Darüber hinaus bildet sich eine steile wandernde Reaktionsfront aus, die den Restumsatz erzielt und die in der katalytischen Schüttung gespeicherte Energie verbraucht. Dieses Verhalten, das ebenfalls von Glöckler et al. experimentell untersucht wurde [3], wird durch detaillierte Simulationen der neuen Reaktorkonfiguration beschrieben. Der Effekt, der aus der angesprochenen Zunahme des Wärmekapazitätsstroms in Strömungsrichtung resultiert, wird in dieser Konfiguration verschärft, sodass der thermische Wirkungsgrad eher schlechter als beim bisher betrachteten Konzept mit katalytisch aktivem Eintrittsbereich ausfällt.

Um diesem Problem zu entgehen, wird das vom M. van Sint Annaland eingeführten Konzept der sog. symmetrischen Fahrweise aufgegriffen [4]. Weist der betrachtete Strömungsumkehrreaktor eine symmetrische Konfiguration mit inerten Ein- und Ablaufbereichen auf, so kann dieser auch symmetrisch betrieben werden. Bei dieser Betriebsweise laufen Produktion und Regeneration zunächst hintereinander in der gleichen Strömungsrichtung ab, wobei in einer zweiten Periode dies in entgegengesetzter Strömungsrichtung erfolgt. Ein wesentlicher Vorteil dieses Konzepts besteht darin, dass für dessen Betrieb nur eine Brennkammer in der Mitte des Festbettes erforderlich ist. Demnach kann auf zwei Brennkammern verzichtet werden, was das Reaktordesign wesentlich vereinfacht.

Die Periodendauer muss so gewählt werden, dass die Energie, die stets in den Randzonen gespeichert wird, für die Aufheizung des Zulaufstroms während der Produktions- und Regenerierphasen ausreicht. Unter Annahme einer idealen thermischen Frontwanderung ohne Dispersion wäre mit diesem Konzept eine vollständige regenerative Wärmerückgewinnung möglich. Die Simulationsergebnisse zeigen jedoch, dass die Temperaturprofile aufgrund der Dispersionseffekte sehr stark abflachen. Für eine optimale Umsetzung des Konzepts ist einerseits eine ausreichende Vorheizung des Zulaufs für die Entstehung der oben genannten stationären Reaktionszone erforderlich. Andererseits ist die Erhaltung hoher Temperaturen am Austritt der katalytischen Schüttung für das Erreichen hoher Umsätze unumgänglich. Wie die Simulationsergebnisse zeigen, kann dies allerdings nur auf Kosten einer reduzierten thermischen Effizienz durch hohe Temperaturen im Ablaufstrom erreicht werden. Jedoch geht die symmetrische Fahrweise mit einer signifikanten Vereinfachung des Reaktors einher, so dass eine experimentelle Validierung und der Ausbau des Konzepts an dieser Stelle empfohlen werden.

Auf Grundlage dieser Analyse erweist sich die von Glöckler vorgeschlagene und im Rahmen dieser Arbeit experimentell validierte Reaktorkonfiguration am besten geeignet. Trotz der eingeschränkten Wärmerückgewinnung können mit diesem Strömungsumkehrreaktorkonzept Vollumsatz und eine hohe thermische Effizienz über einen breiten Lastbereich erreicht werden. Die durch Simulation abgeschätzten Wirkungsgrade liegen im Bereich der thermischen Effizienz, die mit den heutigen optimierten großtechnischen Reformierungstechnologien erreicht wird, weshalb sich das betrachtete Konzept für die dezentrale Herstellung von Wasserstoff besonders auszeichnet.

Abstract

The present work focuses on the further development of the reverse-flow reactor (RFR) concept for the highly endothermic methane steam reforming (MSR) reaction proposed by B. Glöckler et al. in [1, 2]. The RFR is a heat integrated reactor concept that is characterized by the use of the thermal mass of the catalytic packed bed as energy reservoir. In this kind of reactor, regenerative heat recovery is achieved by periodically reversing the direction of the flow. Hence, the energy stored in the outlet section of the packed bed during one period is used to preheat the feed stream during the subsequent period.

This type of operation has been traditionally considered to autothermally run slightly exothermic reactions, e.g. the catalytic purification of industrial exhaust air. Moreover, the excellent heat exchange between the packing and the gas phase offers a huge potential to energetically couple endothermic and exothermic reactions. This is the case of the RFR studied by Glöckler, where the endothermic MSR is coupled with the combustion of a fuel in a regenerative, asymmetric operation mode. This means that the composition of the feed to run the endothermic (production) or the exothermic (regeneration) reactions is switched with every flow reversal.

The reactor is started-up preheating the adiabatic packed bed to temperatures close to 1000 °C, which are required in order to obtain large methane conversions. The MSR reaction takes place during the production period as a self-sharpening endothermic reaction and temperature front that travels in the direction of the flow and consumes the energy stored in the packed bed. Large conversions are reached as long as the front remains in the active packing. When its end is reached, the production step is interrupted and the packed bed is reheated during the subsequent regeneration step.

In order to reheat the packed bed, the direction of the flow is reversed so that regenerative heat integration between production and regeneration is possible. Full heat integration can be ideally achieved whenever the heat fluxes in both steps are identical. During regeneration, the energy consumed by the reforming reaction is supplied to the packed bed through a combustion reaction. As already shown by Glöckler in [1], heat must be released at several positions distributed along the packed bed in order to attain a uniform temperature level. Therefore, the fuel is diluted in a carrier gas that flows through the reactor, while air is introduced through several injection ports, which are confined in inert sections embedded in the active packed bed. The air, the fuel and the carrier gas should mix within the inert sections and react in the catalytic packing downstream. A fraction of the total fuel fed to the reactor is combusted at each of these sections and the energy required to reheat de packed bed is released.

The direct energy release in the packed bed represents an important advantage in comparison to the conventional reforming technology, which generally consists of an externally heated multitubular reactor introduced in radiation furnace. Furthermore, the possibility to take advantage of regenerative energy integration makes it possible to reach large thermal efficiencies, increasing the potential of the reactor concept for the decentralized hydrogen production.

The first experimental validation of the reactor concept was published by Glöckler et al. in [2]. These results, taken as starting point for the current study, describe the occurrence of excess temperatures due to the spontaneous combustion of the fuel at the air injection ports. These temperatures damaged the air injectors and the packing itself and are the cause of an irregular temperature profile at the end of the regeneration step. Thus, the present work focuses in the development of an alternative regeneration strategy, which avoids the problematic mentioned above.

The approach followed inspires in the so called flameless oxidation principle (FLOX®), which is a proven technology applied in industrial burners. These burners are generally operated with excess air and are conceived in order to ensure a homogeneous mixing of the air, the fuel and the combustion off-gases. At sufficiently high temperatures in the combustion chamber, the combustion reaction takes place homogeneously and without formation of visible flames. The proper backmixing of the combustion off-gases with the air and fuel introduced to the combustor is a decisive aspect to attain the flameless combustion regime without occurrence of excess temperatures. According to the literature available, this backmixing can be characterized by the so called recirculation ratio K_V, which represents a relation of the recirculated combustion off-gases and the reactants and should exhibit values above 2.

Adaption of this principle to the RFR implies that the inert sections previously described must be replaced by void sections that are run as flameless combustion chambers during the regeneration step. The design of the chambers must ensure a homogeneous mixing of the air, the fuel and the combustion off-gases. In order to design a chamber that fulfills these requirements, several methodologies have been applied: the flow pattern in the chamber, the mixing behavior of the gases and the chamber performance under reacting conditions has been analyzed based on CFD simulations, visualization experiments using tracer methods and combustion experiments in a mock-up of the combustion chamber.

The concept described above, in which the fuel is supplied diluted in an inert carrier gas, is taken as reference for the development of the combustion chamber. The chamber is operated under fuel-rich conditions ($\lambda < 1$) and the air is injected in such a way that an intense recirculation of the gases in the chamber is achieved. Therefore, the required geometry and alignment of the air injectors as well as the propitious relation between the air and the carrier gas flows are analyzed. This is done based on CFD simulations and visually, using a mock-up of the chamber that is run with water considering the Reynold's similarity of the flows. These studies are followed by combustion experiments in a quartz glass reactor that enable the evaluation of the temperature distribution in the chamber under reacting conditions. In the resulting design of the chamber, the air is introduced through a nozzle head that is radially centered at the outlet of the chamber in the flow direction. Thus, the air is injected in opposite direction to that of the main flow, consisting of the fuel diluted in the carrier gas.

The observations deriving from the investigations mentioned above are consistent and suggest that the design of the air injection nozzles as well as a large ratio between the flow fed through the air nozzles (side stream) and the main stream are decisive parameters to establish flameless combustion regimes. If large side-to-main flow ratios cannot be maintained during reverse-flow operation, formation of flames may occur. However, the flames are locally confined within the void section of the chamber, so that the resulting temperature peaks do not represent a risk for the stability of the catalyst or the construction materials.

The ignition behavior of the chamber is also an important aspect to be considered under reverse-flow operation. Since the packed bed and the combustion chambers are cooled down during the production step, their temperature at the beginning of the regeneration step is below the ignition temperature of the homogeneous combustion reactions. Thus, the air and the fuel are homogeneously mixed in the chamber and react in the catalytic packing downstream. The combustion reaction takes place in the catalyst until the temperature in the chamber exceeds the ignition temperature of the gas mixture and homogeneous combustion occurs. According to this description of the ignition behavior, a considerable ignition delay or induction time must be accounted for. In spite thereof, given the homogeneity of the mixture containing the fuel and the combustion air, no excess temperatures occur during this process. Moreover, a controlled, uniform combustion with almost constant adiabatic temperature increase can be run also under dynamic conditions.

The operation of the combustion chamber under fuel-rich conditions exhibits a strong sensitivity towards the fuel if light hydrocarbons are used. Experimental results corroborate that partial oxidation takes place. Part of the fuel reacts at the inlet zone of the active packing causing a temperature increase that is followed by a temperature decrease caused by the cooling effect of subsequent endothermic reactions that the unreacted fuel undergoes in the active packing downstream. This observed behavior helps to describe the irregular temperature profiles at the end of the regeneration step when using methane containing fuels, as in the experimental validation of the concept reported by Glöckler.

With the aim to run the RFR in a decentralized hydrogen production facility accounting for energy and material integration, the off-gases of a pressure swing adsorption (PSA) for the purification of the produced hydrogen are to be used as fuel during regeneration. Since this gas contains significant amounts of methane, an alternative regeneration strategy that accounts for the suppression of the methane reforming reactions during the reheating step must be developed. A straightforward approach would consist of introducing air in the main flow and injecting the fuel in each of the combustion chambers, so that these are run under air-rich conditions ($\lambda > 1$). However, the oxidation of the catalyst could prove disadvantageous during the subsequent production step. Thus, a reheating strategy based on stoichiometric combustion ($\lambda = 1$) in each chamber is pursued.

Based on this concept, the reactor is fed with an inert gas, whereas the air and the fuel are supplied stoichiometrically and separated from each other in every combustion chamber. As already described, the proper mixing of the three streams is decisive to operate the chamber under flameless conditions. The experimental determination of the required parameters and the study of the behaviour under combustion conditions is therefore performed in a metallic combustion chamber. In contrast to the chamber studied so far, the current design accounts for an additional lateral fuel supply positioned at half of its height.

The results obtained confirm the advantages deriving from the stoichiometric operating conditions. These advantages refer not only to the successful suppression of undesired endothermic after-reactions during combustion, but also to the simplification of the control strategy of the chamber during operation under flow reversal. Based on the experimental data and on the analysis of the temperature homogeneity and the fuel conversion, it is possible to define an operation region in which large fuel conversions ($> 90\%$) and small temperature differences within the chamber can be achieved. This operation region, however, is strongly limited by the concept followed to introduce the fuel separately into the chamber.

The experimental observations are validated using CFD simulations. The results indicate that the dilution of the fuel in the carrier gas plays a decisive role in order to ensure a homogeneous distribution of the combustion reaction in the gas bulk. Accordingly, flame suppression and a homogeneous temperature distribution in the chamber can be rather achieved with the original chamber design, in which the fuel is diluted in the carrier gas, than in the current concept, in which the fuel is introduced as a side-feed. However, as already described and experimentally verified, the formation of flame fronts occur in the void of the chamber, so that no catalyst and internals damage must be accounted for.

The regeneration concept is verified under reverse-flow conditions taking the findings so far into consideration. Three of the described combustion chambers are embedded in a steel reactor conceived for this purpose. The RFR is operated at reforming loads between 2 and $5\,kW_{LHV,H_2}$. During regeneration, either H_2 or PSA off-gas can be used as fuel. The fuel is supplied together with combustion air in stoichiometric proportions to each of the chambers based on a control strategy that enables to heat them up to a constant temperature set point.

The operating parameters for reverse-flow operation are adjusted in order to efficiently recover the energy contained in the effluents, while the duration of the production and regeneration steps is chosen to be equal. The mass flow during each of the steps is adjusted so that the heat flux in both directions is equivalent. This has an effect on the attainable heat recovery, which can be derived from the temperature difference between the reactor feed and its effluents. Thus, the reactor is operated in such a way that the feed exhibits temperatures around 150 °C, whereas the temperature of the effluents should not exceed 300 °C.

The results obtained under reverse-flow operation corroborate that an efficient reheating at a high temperature level, avoiding the occurrence of excess temperatures is possible. In contrast to the experimental results described by Glöckler et al. in [2], a uniform temperature profile at about 900 °C can be established at the end of the regeneration step. Even under operating conditions at which large temperature differences in the chamber cannot be successfully avoided, these differences can be immediately smoothened after begin of the production period, given the low heat capacity of the combustion chamber.

The performance of the current reactor concept proves to be reproducible and independent of load variations within the design operation range. Conversions above 90% are achieved at moderate reforming loads, corresponding to the equilibrium conversion at the conditions at the outlet of the catalytic packing. The behavior of the temperature profiles at the reactor boundaries is a good indicator for the satisfactory heat recovery. In spite thereof, thermal efficiencies of only 40 to 50% are obtained, due to the influence of heat losses in the experimental setup and due to an increase of the heat flux in the direction of the flow during both periods. As discussed later on, the increase of the heat flux is dependent on the physical properties of the flow and on the fuel used for regeneration. The proper adjustment of the operation parameters, i.e. the mass flow during production and regeneration, contributes to partially compensate this effect. This is shown with help of detailed simulations, which are the tool chosen to perform a detailed analysis of the system and for the study of conceptual variations of the RFR concept.

The mathematical model developed for this purpose describes the packed bed as a multiphase, plug-flow reactor with axial dispersion. The combustion chambers are modeled as ideally mixed stirred reactors, assuming total conversion during the regeneration step. The simulation tool used is DIANA, developed at the Max Planck Institute for dynamics of complex technical systems.

As already introduced, an optimal heat recovery assumes constant and equal heat fluxes in both flow directions. However, the heat flux over the reactor length increases about 35% with the heat capacity of the gas during the production step as a result of the reforming reaction. On the contrary, the heat flux during the regeneration step increases due to the supply of fuel and combustion air along the reactor up to 15% if hydrogen is used as fuel, and 30% in case that PSA off-gas is used. Accordingly, the heat flux of the reactor effluents is always larger than the one of the feed, thus limiting the energy recovery. Moreover, the heat flux of the flow during the production step is almost twice the heat flux of the regeneration gas flow. Hence, a regeneration-to-production mass flow ratio of at least 2 is required during reverse-flow operation. Based on the simulation results, the largest thermal efficiencies by full methane conversion can be achieved at mass flow ratios of around 2.5. Although the efficiencies attained are beyond 80%, the results suggest that these cannot be further enhanced and that the difference of the production and regeneration heat fluxes cannot be compensated by means of adjustment of the mass flow ratio or the switching frequency between periods.

The simulation tools, however, enable the study of alternative reactor and packed bed configurations. In the current reactor concept, the inlet section during the production step is filled with active packing, while the outlet zone is filled with inert packing in order to avoid back-conversion when the effluents are cooled down. Although the reactants are preheated in the inlet section, the temperatures in this zone of the packing are not sufficiently high to substantially contribute to the total conversion of the system. Hence, the proposition made by Glöckler et al. in [3] is adopted and the active section at the reactor inlet is replaced by inert packing. If the energy stored in this section during the regeneration step is sufficient to preheat the reforming reactants to a high temperature, a stationary reaction zone establishes as soon as the hot gases enter the catalytic bed. This reaction zone is characterized by a steep temperature decrease due to the partial conversion of the reforming reactants. Additionally, a steep moving reaction front where the remaining reactants are fully converted travels in the direction of the flow and consumes the energy stored in the active packing. This behavior, which has been experimentally verified and reported by Glöckler et al. [3], is also described based on detailed simulations. However, the results indicate that the negative effects caused by the heat flux increase in the direction of the flow are intensified with this reactor configuration. Thus, the thermal efficiencies achieved are lower than in case of the reactor with an active inlet section during the production period.

In order to avoid this problematic, the so called symmetric operation switching pattern proposed by M. van Sint Annaland is taken into consideration [4]. As long as the packed bed is symmetrically structured, with inert inlet and outlet sections, the reactor can be operated symmetrically. Accordingly, production and regeneration are performed in the same flow direction in a first period, whereas this is repeated in the opposite direction during the subsequent period. An essential advantage of this concept is the possibility to eliminate two of the three combustion chambers, so that the reactor can be run with a single chamber embedded in the middle of the packed bed.

The duration of the different steps must be adjusted in order to store enough energy at the reactor boundaries, so that the feed can be sufficiently preheated during the consecutive production and regeneration steps. This operation concept enables an optimal energy recovery if the dispersion effects in the evolution of the thermal front can be neglected. However, the detailed simulation results show that this is not the case and that the dispersive effects cause the temperature profiles to strongly flatten during operation.

The optimal realization of the concept requires preheating the reactants during the production step to a high temperature, so that the stationary reaction zone described above can be formed. At the same time, the temperature at the outlet of the catalytic zone needs to be kept at a high level, in order to reach high conversions. According to the simulation results, this is only possible at the expense of reduced thermal efficiencies and large effluent temperatures. Nevertheless, the simplification of the reactor design through elimination of two combustion chambers is an aspect that encourages an experimental validation and the further development of the concept.

Based on this analysis and taking into account the attainable thermal efficiencies of the different configurations studied in this work, the RFR concept originally proposed by Glöckler represents the most appropriate reactor configuration. The regeneration strategy developed in this work to overcome the technical limitations of the original concept has been experimentally validated. Furthermore, large thermal efficiencies and conversions can be attained with this concept over a wide operation range. The efficiencies estimated based on detailed simulations are in the same range than those achieved in the optimized state-of-the-art reforming technologies. Thus, it can be stated that the RFR concept originally analyzed is especially suited for the production of hydrogen in decentralized facilities.

Chapter 1

Introduction

1.1 Autothermal reactor concepts for endothermic high-temperature syntheses

Heat integrated reactors represent an important subcategory within the so called multifunctional reactors that in turn, constitute a core element in the philosophy of process intensification. Regulation of temperature, together with concentration or activity profiles along a reactor is one of the pillars for the optimization of its performance. The use of the term *multifunctional* or *extended reactor* was already used by Agar et al. in [5] referring to a broad class of novel reactor configurations for heterogeneously catalyzed reactions. The latter combine mass and energy transport in such a way, that the operation conditions can be optimized for a specific reaction system.

Within the class of multifunctional reactors, heat integrated or heat exchanger reactors play an important role, provided that some configurations represent well established and widely used concepts in industry. Multitubular fixed bed reactors or adiabatic reactors with intercooling/heating sections are probably the most representative examples that aim the establishment of an optimal temperature profile to perform a given reaction [6]. Whenever the overall reaction system considered turns to be slightly exothermic, recovery of the energy contained in the effluents enables the so called autothermal operation, meaning that no external supply is needed to sustain operation in an ignited state. Coupling of an adiabatic reactor with a heat exchanger or recuperator is probably the best-known and straightforward concept of such a heat integrated system.

The realization of an endothermic synthesis without the need of external heat supply through appropriate spatial or temporal coupling with an exothermic reaction is also considered autothermal operation. Agar describes such a case in [7], taking hydrogen cyanide production by reaction of methane with ammonia as example. However, as pointed out later on by Kolios in [8], an endothermic reaction can only be run if the required energy is provided at the reaction temperature. This apparently trivial consideration turns into a crucial aspect when considering high-temperature syntheses. Heat transfer limitations in conventional heat integrated reactor concepts pose a great technological and engineering challenge that becomes particularly acute at high temperatures. This problematic defines the general frame of the current work, that tackles the subject matter taking hydrogen or synthesis gas production via methane steam reforming (MSR) as example, given its large industrial interest [9, 10].

Kolios et al. reviewed in [8, 11, 12] a row of reactor concepts based on both *recuperative* and *regenerative* heat exchange, to autothermally perform not only weakly exothermic reactions but also to energetically couple the endothermic MSR with an exothermic combustion in a single unit. In all cases, the different configurations strive heat exchange between the hot effluents and the cold reactants, ensuring low temperatures at the reactor boundaries and thus, making these systems appropriate for decentralized applications.

Under *recuperative* operation is understood the thermal coupling of two neighbouring, spacially separated compartments, in which the exothermic, the endothermic or a combination of both reactions occur respectively. Several reactor concepts accounting for counter- and co-current operation depending on the direction of the flow in each of the reaction compartments have been extensively reported in the literature. Frauhammer et al. [13] and Kolios et al. [14, 15] analyzed the behaviour of counter-current operation and discussed the several difficulties to extend the zone at which both reactions can be effectively coupled. These observations were corroborated by Ramaswamy et al. [16], in an analysis of both operation modes (co- und counter-current) that pays special attention to the transient reactor behavior (e.g. during the start-up procedure). Kolios et al. proposed in [11, 12, 17] an alternative concept based on the combination of a co-currently operated reaction section and counter-current recuperative heat exchangers at the reactor boundaries. The analysis and successful experimental proof-of-concept was reported by Gritsch et al. [18–20].

Besides the recuperative systems mentioned above, *regenerative* reactor concepts represent an interesting alternative that uses the thermal mass of the packed bed as energy reservoir. Heat transfer limitations can be herewith efficiently minimized, given the large heat transfer coefficients between gas and solid phases in packed beds. A promising reactor concept exhibiting large potential for the coupling of exo- and endothermic reactions at high temperature levels is the so called *Reverse-flow reactor*, abreviated as RFR in the following. The characteristic behaviour of the latter has been theoretically and experimentally analyzed and extensively reported for weakly exothermic reactions. For instance Nieken [21–23], Züfle [24], Khinast [25, 26] and Aubé [27] reported on the catalytic purification of industrial exhaust air, whereas Bunimovich also considered the treatment of exhaust gases from natural gas fueled diesel engines [28]. Moreover, autothermal RFR operation resulting from the simultaneous combination of an endothermic and an exothermic reaction has been reported by several authors. Systems such as synthesis gas production by catalytic partial oxidation of methane [29–31] or styrene synthesis by dehydrogenation of ethylbenzene with introduction of superheated steam along the reactor length [32] are some of the examples discussed in an extended overwiew of processes with industrial relevance given by Matros et al. [33].

A common characteristic of the systems above mentioned is their *symmetric* mode of operation. The latter implies that the feed temperature and compositions are kept constant while flow reversal is performed. However, focus of this work is the operation of the RFR in the so called *asymmetric* mode in which the reactor is operated cyclically with alternating endothermic and exothermic steps. During the endothermic step, the reacting mixture is fed to the preheated packed bed and the reaction takes place sustained by the energy stored in the packing, which cools down. In a subsequent step, the energy consumed in the endothermic step is *regenerated* by reheating the packed bed with a hot gas flow or more specifically, performing an exothermic reaction (combustion) and transporting the energy released with the product gases. At the same time, flow reversal is performed in order to trap the high temperature region in the center of the packed bed, maintain low effluent temperatures and thus, improve the energetic efficiency of the system.

This form of operation has been considered, for example, by Kolios et al. for the styrene synthesis combining ethylbenzene dehydrogenation and combustion of recovered hydrogen [34], by Van Sint Annaland et al. for light parafine dehydrogenation and combustion of hydrocarbons (and carbon deposits) [4, 35, 36], as well as for MSR and combustion of hydrocarbons by several authors [8, 37, 38]. A major issue in all cases referred is the proper spatial adjustment of the endothermic and exothermic reaction zones. If the reaction zones do not overlap, a region establishes in which energy is stored during each regeneration and not consumed during the following endothermic step, leading to the formation of undesirable temperature peaks. The latter might, amongst others, irreversibly damage the catalyst. The maximum temperatures need to be kept within a reasonable range by dilution [35] or dedicated preheating of the process streams [38]. However, an elegant approach to enhance the overlapping of the reaction zones is to axially distribute the energy supply during the regeneration period, in analogy to the proposal from Snyder and Subramaniam for styrene synthesis, in order to attain an almost isothermal operation mode [32].

This concept, adopted and developed by Glöckler et al. to a great extent for MSR [1, 39, 40], sets the fundamentals of the present work.

1.2 Motivation and content overview

The potential of the RFR concept to efficiently perform the highly endothermic MSR reaction with spatially distributed heat supply during the regeneration step has been thoroughfully discussed by Glöckler et al. on a theoretical basis [1, 39]. However, the experimental proof-of-concept presented by the same authors [40, 41] revealed that the phenomenon of excess temperatures in the position of the heat sources could be associated to the simultaneous occurrence of homogeneous and heterogeneous reactions, posing a technological hurdle when striving an efficient, uniform reheating of the packed bed at a high temperature level based on combustion reactions.

At present, the so called *flameless oxidation* (FLOX®) technology, primarily developed for heating systems in industrial furnaces, is gaining on attention and significantly expanding its application areas [42–45]. This technology, based on the effective supression of flames during combustion to attain energetic efficient combustion regimes with mild temperatures and low emissions, represents a refined alternative to provide the RFR with energy at high temperature levels in a controlled manner. However, the adaption of this novel technology to enable its embedment in a packed bed and the successful integration in a dynamic system operated under flow reversal are accompanied by a row of technologic challenges that motivate the development reported in the present work. Thus, optimization of the concept proposed by Glöckler et al. with special focus on the regeneration step is the central aspect of the current study, which not only addresses how to appropriately design and operate a FLOX® chamber to efficiently reheat a packed bed, but also considers the embedment of this technology in a RFR.

The appropriate reactor design and the establishment of a suitable operation strategy based on a preferably simple control strategy are the core issues of this work, which are tackled and reported based on the following structure. The state of the art in the production of hydrogen by MSR is briefly reviewed in chapter 2. Furthermore, it focuses on the autothermal, regenerative asymmetric operation mode and summarizes the features of the system that make possible this form of operation. The research work conducted by Glöckler et al. is reviewed and the challenges resulting thereof, which motivate the content of this work, are exposed at the same time that the process constraints for the further study are concretized.

Chapter 3 deals with the fundamentals of the FLOX® principle and the development of a combustion chamber enabling its use to reheat a packed-bed reactor. The proof-of-concept and the definition of an appropriate operation range is discussed in detail in chapter 4.

Chapter 5 maintains the experimental character, focusing however on the performance of a RFR under periodic operation. Therefor, the combustion chamber and its behaviour already described in chapters 3 and 4 is tested under flow reversal conditions. The performance and limitations of the system are discussed based on experimental results. The latter are qualitatively corroborated and analyzed based on a mathematical model of the system and the simulation results generated with it. The model itself is described in chapter 6, whereas the simulation results are thoroughly discussed in chapter 7. Moreover, the modelling tool is used to explore altenative operation modes and provide the basis for further studies. Concluding remarks and an outlook for the current work is given at the end of the same chapter.

Chapter 2

Autothermal Reverse-Flow Reformer

Autothermal reactor concepts based on regenerative heat exchange are characterized by an excellent heat transfer between gas and solid packing. Furthermore, scalability and the simplicity of their design are aspects that clearly increase their attractiveness.

Regenerative concepts, energetically coupling endothermic and exothermic reactions, have a long industrial tradition in case of light paraffin dehydrogenation. This is the case, for instance, of the Houdry Catofin process from ABB Lummus. During dehydrogenation, the catalyst exhibits a rapid deactivation due to coke deposition over its surface. Thus, the need to periodically regenerate its activity through combustion of the coke deposits enables the recovery of the energy released. However, according to Stitt [46], whenever catalyst deactivation does not affect the process, the economic potential of this reactor concept compared to conventional designs is questionable. This is the case, for example, of catalytic reforming of natural gas on an industrial scale, provided that it represents a well-established mature technology.

Nevertheless, as discussed in the following, conventional MSR technology poses several challenges when considering decentralized production of hydrogen. This trend arises from the need to establish a robust hydrogen supply infrastructure and to lay the foundations of a hydrogen based economy using the latter as a sustainable energy carrier. The RFR concept does not only pose great advantages for this purpose, but the nature of the reaction system enables exploiting to a great extent the attributes of asymmetric operation.

2.1 Methane steam reforming

Production of synthesis gas stays at the beginning of the value chain in the chemical industry, the H_2/CO ratio depending upon the technology used for its production. Hence, the latter is generally chosen in order to match the requirements of the downstream processes. For instance, catalytic partial oxidation of methane (CPOM) or autothermal reforming (ATR) provide H_2/CO ratios in the order of 2, convenient for Fischer-Tropsch or methanol synthesis. Conversely, MSR provides larger ratios in the order of 3, becoming the technology of choice when hydrogen is the desired product.

Whereas CPOM or ATR are implicitely run in an autothermal way, conventional MSR is traditionally run on a recuperative basis. Conventional tubular reformers consist of externally heated tubes placed in a radiation furnace. The reforming reaction itself is run at temperatures in the range of 800 -900 °C and pressures up to 15 -30 bar. In order to overcome heat transfer limitations and enable effective energy transfer, large temperature gradients between the heating (outside the reformer tubes) and the reaction medium (flowing through the packed bed) are required. In spite of the large overall thermal efficiencies of reformer and convection zones (up to 92%), only about 50% of the total energy supplied is transferred to the tubes, whereas the rest leaves the furnace in form of hot flue gases. This energy and that contained in the product flow that leaves the reformer at temperatures around 850 °C, needs to be recovered in order to achieve the desired energetic efficiencies [10].

Whenever energy export to downstream processes is not possible, e.g. for decentralized applications, alternative systems need to be explored. As briefly discussed in section 2.2, the efforts to energetically optimize the MSR process have been mostly made based on recuperative concepts. The energy exchange efficiency in such concepts, however, is far from that achievable in regenerative concepts, such as the RFR discussed in section 2.3. Accordingly, larger reactor volumes are required in order to achieve comparable energy integration degrees.

2.1.1 Reaction system

The overall raction system describing the MSR process consists of the endothermic methane steam reforming and exothermic combustion reactions, responsible for the required energy supply.

Methane steam reforming: Within all possible hydrocarbons, natural gas (NG) currently represents the most economic feedstock for catalytic steam reforming [10]. Being methane its main component, MSR is properly described considering the highly endothermic, equilibrium limited reforming (2.1) and direct reforming (2.3) reactions between methane and steam and the slightly exothermic water-gas shift (WGS) reaction (2.2).

$$\text{Reforming:} \quad CH_4 + H_2O \rightleftharpoons CO + 3H_2 \qquad \Delta h_{r,1} = +206.28\,\text{kJ/mol} \tag{2.1}$$

$$\text{Water-gas shift:} \quad CO + H_2O \rightleftharpoons CO_2 + H_2 \qquad \Delta h_{r,2} = -41.16\,\text{kJ/mol} \tag{2.2}$$

$$\text{Direct-Reforming:} \quad CH_4 + 2H_2O \rightleftharpoons CO_2 + 4H_2 \qquad \Delta h_{r,3} = +165.12\,\text{kJ/mol} \tag{2.3}$$

The approximate composition of the reformer effluent is determined by the simultaneous equilibrium of reactions (2.1) and (2.2). In order to recover the maximal amount of hydrogen possible, the mixture leaving the reformer may be further reacted in order to convert CO to CO_2. Conversely to the reforming reactions, the WGS reaction as written in equation (2.2) is favoured at moderate temperatures. Thus, a so called WGS stage or a combination of several stages at different temperature levels (high- and low-temperature WGS) is generally installed downstream of the reformer [9, 10]. Hydrogen can be recovered, for example, by passing the mixture through a pressure swing adsorption (PSA) unit.

In the current study, MSR is performed at a steam/carbon (S:C) ratio of 3. Accordingly, consideration of soot formation due to thermal cracking (2.4), disproportionation (Boudouard) (2.5) or CO reduction (2.6) reactions can be omitted [9].

$$\text{Thermal decomposition}: \quad CH_4 \rightarrow C + 2H_2 \qquad \Delta h_{r,4} = +74.9\,\text{kJ/mol} \tag{2.4}$$

$$\text{Boudouard reaction}: \quad 2CO \rightleftharpoons C + CO_2 \qquad \Delta h_{r,5} = -172.54\,\text{kJ/mol} \tag{2.5}$$

$$\text{CO reduction}: \quad CO + H_2 \rightarrow C + H_2O \qquad \Delta h_{r,6} = -150.0\,\text{kJ/mol} \tag{2.6}$$

Combustion: The total energy consumed by the endothermic MSR is supplied to the system by exothermic combustion reactions. Whereas in conventional steam reformers, combustion reactions take place in a firing section, in the RFR energy is released in the packed bed during the regeneration period. The reaction system depends on the fuel used. Most of the chemical species available within the reforming facility are combustible and can therefore be used as fuel. However, combustion of pure H_2 or CH_4 can be considered as productivity losses. Thus, the use of PSA off-gases as fuel is generally preferred. In an ideal autothermal operation regime, energy released by combustion of the PSA off-gases should suffice to sustain the endothermic reforming period (cf. appendix E).

The composition of the PSA off-gases depends strongly on the operation conditions of the reformer and the PSA unit. CH_4, H_2 and CO contained in the gas mixture are the combustible fractions. Accordingly, under assumption of complete, stoichiometric reaction, the reactions governing the combustion system can be described as follows:

$$CH_4 + 2O_2 \rightarrow CO_2 + 2H_2O \qquad \Delta h^0_{r,7} = -802\,\text{kJ/mol} \tag{2.7}$$

$$CO + 0.5O_2 \rightarrow CO_2 \qquad \Delta h^0_{r,8} = -283\,\text{kJ/mol} \tag{2.8}$$

$$H_2 + 0.5O_2 \rightarrow H_2O \qquad \Delta h^0_{r,9} = -242\,\text{kJ/mol} \tag{2.9}$$

The current study uses two different types of fuel as reference: pure H_2 is used due to its ignitability and simplicity (H_2O is the only product), whereas synthetic PSA off-gas with a composition according to table 2.1 is used as reference of the available fuel under technical conditions.

Table 2.1: PSA off-gas molar composition in conventional MSR facilities estimated according to industrial data [10, 47].

	H_2	N_2	CO	CO_2	CH_4
y_j	34,67%	1,01%	13,64%	38,08%	12,61%

2.2 Energy efficient reactor concepts for MSR - State of the art

Improvement of reforming technology is, as already mentioned, mainly performed in the field of recuperative heat exchanger reactors. Optimization potential of established MSR technologies can be achieved through improvement of the endothermic reaction performance and of the combustion process efficiency. The first aspect is closely related to catalyst development. The geometric design is a core tuning parameter in order to enhance the catalyst specific surface, reduce pressure drop and intensify the heat flux to the catalyst [9]. Heat transfer enhancement and optimization of heat flux profiles in conventional fired reformers has indeed proven to have the largest effect on the operation performance based on several theoretical studies [48–50]. Accordingly, Topsøe has made use of material improvements making it possible to reduce the diameter of the reformer tubes, their thickness and thus, increase the heat flux in the so called High Flux Steam Reforming (HFR), designed for capacities of 200.000 $Nm^3/h\,H_2$ [51]. This improvement, however, affects primarily the dimensions of the furnaces and reduces the investment costs but does not imply significant improvement in the reformer thermal efficiency.

In order to improve the amount of energy transferred to the process gas and reduce the steam export accordingly, the so called convective reforming, which recovers the energy contained in the reformer effluents or in the flue gas of the firing section, represents an interesting alternative. However, the fact that convection instead of radiation becomes the primary heat transfer mechanism implies a reduction of the heat fluxes, resulting in still large reactor volumes [52]. The straightforward realisation of convective reforming uses the energy contained in the reactor effluent according to the Gas heated reformer (GHR) developed by the ICI Ltd. or the Haldor Topsøe exchange reformer (HTER) technologies.

Both are generally implemented in combination with conventional MSR or ATR enabling capacity increases in the order of $20-30\%$ and reduction of steam production [52–54].

A further improvement in order to fully suppress steam export is the Haldor Topsøe convection reformer (HTCR). The latter not only recuperates counter-currently the energy contained in the process effluent with a concentric reformer tube design, but it also uses the energy contained in the flue gas of the firing section to convectively heat the outer wall of the reformer tubes [54, 55]. Typical capacity ranges for this technology are up to $30.000\,\mathrm{Nm^3/h\,H_2}$. However, despite compactness of design and achievement of great energy integration (about 80% of the heat release in the burner is used in the process so that steam export can be avoided), the main drawback are implicit transport limitations of convective heat exchange, requiring still relatively large reactor volumes.

In a somewhat lower capacity range ($50-400\,\mathrm{Nm^3/h\,H_2}$), WS Reformer has developed an interesting concept. In the so called FLOX®-Reformer Modular, focus of the optimization are both the reformer tubes and the burner section. The design of the reformer tubes resembles that of the HTCR concept, concieved in order to recover the energy of the product gases. However, the peculiarity relies in the fact that the heat exchanger to generate the process steam is integrated in the reformer tube, contributing to cool down the product gases and providing self-adaptivity to the operation conditions when varying the reforming load. Contrary to HTCR, the reformer tubes are located in the combustion chamber and thus, primarily heated by radiation. Since the combustion section is designed and operated based on FLOX® technology (cf. section 3.1), high energetic efficiency and compactness of design can be achieved. In this case, radiative heat transfer is the limiting factor regarding the minimal unit dimensions for a given productivity [56].

Besides the above mentioned concepts, a row of recuperative reactor concepts in the micrometer and millimeter range for low capacity applications, exhibiting large volume specific surfaces and enhanced heat and mass transport characteristics, have been extensively reported in the literature. Gritsch reviewed some of the most relevant ones in [19]. However, looking at larger production capacities ($>400\,\mathrm{Nm^3/h\,H_2}$), there is indeed a need to develop further decentral autothermal reactors that overcome the previously discussed heat transfer limitations. The regenerative, asymmetrically operated reverse-flow reformer discussed in the following section represents a promising alternative.

2.3 Reverse-flow reformer

Major advantage of the RFR is the implicit minimization of heat transfer limitations to which recuperative concepts are generally subject. This translates in a higher energetic efficiency of the reactor system and consequently, of the whole hydrogen production facility. Preliminary estimations suggest that due to the high energy integration in the RFR, the heating value of the off-gases leaving a PSA unit operating under industrially relevant conditions might suffice to enable an energetically self-sustaining operation of a whole decentralized hydrogen production unit (cf. appendix E). Since the production costs of a MSR facility are manily driven by the raw materials costs [57], being able to minimize the use of NG as fuel represents a key aspect when considering the potential of the RFR as alternative for this kind of operation.

The operation principle of the RFR is shown in figure 2.1. Since operated in an asymmetrical manner, flow reversal is performed at the time that the feed composition changes. Thus, the endothermic MSR and the exothermic combustion reactions occur temporally distributed in the same space. The switching pattern between both reactions is such, that an alternating operation mode consisting of an endothermic or production semicycle followed by an exothermic or regeneration one is established. Hence, the energy consumed by the endothermic reforming reaction is introduced in the subsequent semicycle. Spatial overlapping of the endothermic and exothermic reaction zones is essential for the optimal energetic coupling. As discussed in the following, the temperature profiles evolving during both operation periods are decisive when defining an optimal reactor design and its operation.

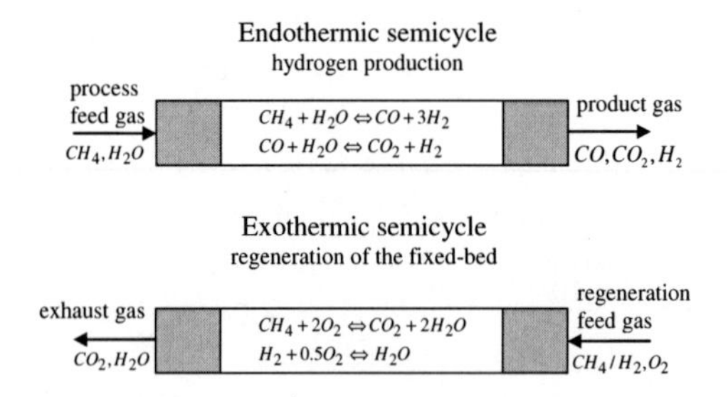

Figure 2.1: Sketch of the regenerative RFR under asymmetric operation, coupling the endothermic MSR with a combustion reaction.

2.3.1 Fundamentals of the operation principle

The fundamentals of the RFR concept have been thoroughly studied by Glöckler and collaborators [1, 39]. The operation principle is strongly dependent on the transient behaviour of the reaction front developing during the production period. The characteristics of this front, in turn, determine the optimal regeneration strategy. An exhaustive analysis of several forms of reheating the packed bed based on a simplified model of the RFR has been reported by Glöckler in [39], whereas the essential characteristics defining the concept studied in the current work have been very well synthesized in [1]. The latter is referred for a detailed description of the operation principle.

Production period

Figure 2.2 appropriately describes the characteristics of an ideal production cycle. As it can be seen from the sketch on the top, the original reactor concept consists of a catalytic packed bed followed by an inert bed at the reactor outlet. The temperature profile at the beginning of the production cycle is characterized by a uniform temperature plateau over the reactor length, whereas the boundaries exhibit low temperatures in accordance to the target of heat export minimization. During the production semicycle, the reactants (CH_4 and steam) are fed into the reactor at a moderate temperature, slightly above the dew point. They are regeneratively heated up at the left reactor boundary while cooling down the packed bed. At sufficiently high temperatures, the endothermic reaction takes place sustained by the energy stored in the packed bed. An endothermic reaction front travelling in the direction of the flow evolves, consuming the energy stored and cooling the packed bed.

Provided that reaction kinetics are fast enough, it can be assumed that thermodynamic equilibrium is attained at each position along the active reactor packing. Thus, a conversion profile evolves according to the temperature profile as depicted in the bottom plot of figure 2.2. The temperature of the gas leaving the catalytic zone determines the conversion attained during the whole period and is to be chosen according to the desired productivity. The reactor effluents are cooled down in the inert section at the reactor outlet, filled with inert packing in order to suppress reverse reactions, i.e. methanation, as already practiced by van Sint Annaland et al. in [35, 36]. As it can be extracted from the evolution of the temperature and conversion profiles, the production period ends as soon as the reactor front reaches the end of the catalytic zone or, alternatively, the temperature of the reactor outlet increases beyond a desired level.

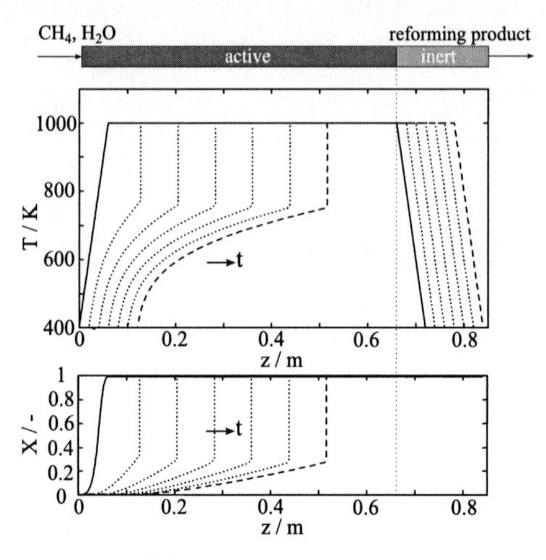

Figure 2.2: Temperature and conversion profiles over the reactor length during the production period based on a simplified model [1].

The form of the endothermic reaction front is a key aspect for the performance of the RFR for two reasons. The formation of a stable, steep front guarantees an optimal use of the energy stored in the packed bed, enabling large space-time-yields and a constant conversion during the production period if uniformity of the temperature plateau is given. Furthermore, the form of the front directly affects the appropriate regeneration strategy, as discussed later on. Focusing on the first aspect, the characteristics of the endothermic reaction front can be discussed based on equation (2.10), which results from an energy balance of the system and describes the dependency of the front velocity on temperature [1].

$$w(T) = \underbrace{\frac{(\dot{m} \cdot c_p)_g}{(1-\varepsilon) \cdot (\rho \cdot c)_s}}_{w_{therm}} \cdot \left(1 + \left|\Delta T^{ad}_{endo}\right| \cdot \frac{dX}{dT}\right) \tag{2.10}$$

Accordingly, figure 2.3 a shows that in a non-dispersive system and under assumption of constant physical parameters, the velocity of the endothermic front w consists of a constant thermal or convective term w_{therm} and a further contribution deriving from the energy consumption by the endothermic reaction. Since conversion is strongly dependent on temperature, the resulting velocity profile exhibits a maximum at the temperature at which the conversion change with temperature is largest, named T_{crit} in the following and fulfilling

$$\left.\frac{dX}{dT}\right|_{T_{crit}} = max \tag{2.11}$$

with $X(T)$ being a function that describes the dependency of methane equilibrium conversion on temperature [3]. Interestingly, as depicted in figures b and c, if the initial temperature plateau is equal or below T_{crit}, the front develops following a dispersive pattern. On the contrary, at initial temperatures beyond T_{crit}, the front behaves as a self-sharpening front, so that the resulting behaviour comprises both a dispersive and a compressive region.

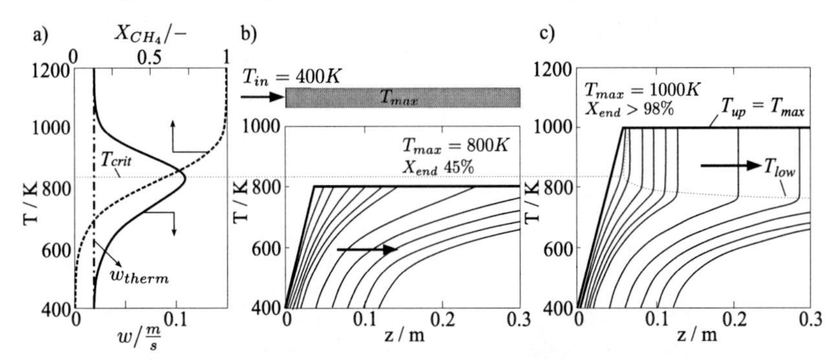

Figure 2.3: Plots summarizing the formation of an endothermic, self-sharpening front at temperatures beyond that corresponding to the maximal migration velocity. a) Velocity profile according to equation (2.10) and methane conversion as a function of temperature. b) and c) Formation and migration of the endothermic reaction front with initial linear temperature profiles and different plateau temperatures. Extracted and adapted from [1, 2].

As reported in [1], the height of the front extends over a temperature region between T_{up} and T_{low} (s. fig. 2.3 c), being T_{low} the temperature at which the maximum velocity of the dispersive front equals the velocity of the shock front. This temperature, which depends on T_{up} or what is the same, the preheating temperature of the bed, can be recursively estimated from

$$\frac{dX}{dT} = \frac{X(T_{up}) - X(T_{low})}{T_{up} - T_{low}} \tag{2.12}$$

The existence of the shock wave described above, crucial for the design of the RFR and for an efficient energy usage, has been experimentally validated and reported in [2, 3, 58, 59]. However, for the sake of completeness, relevant results showing the formation and movement of a sharp front are shown in figure 2.4 and will be briefly discussed.

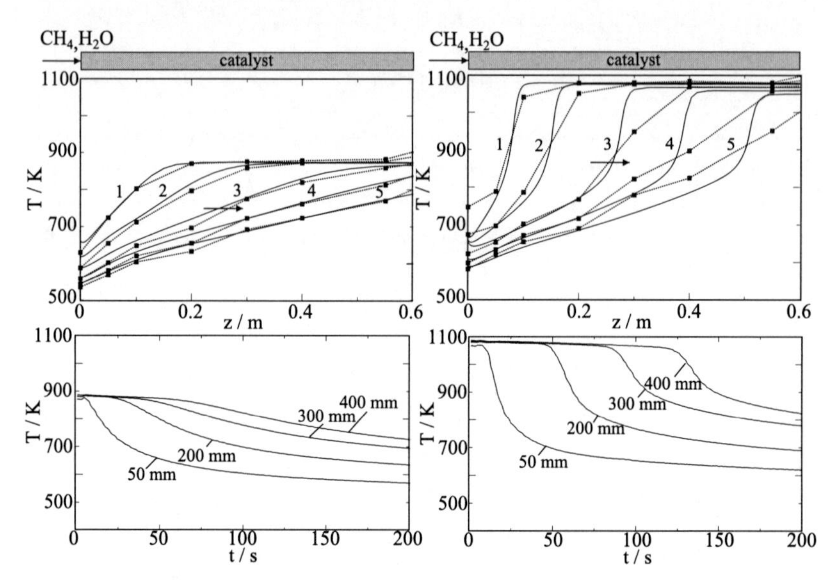

Figure 2.4: Temperature evolution during methane steam reforming in a catalyst bed uniformly preheated at 600 (left) and 800 °C (right). Top: measured (points) and simulated (straight lines) temperatures along the packed bed at 25, 50, 100, 150 and 200 s after reaction beginning (profiles 1 to 5 respectively). Bottom: measured temperature evolution over time at the specified positions. Feed: 7.8 slm CH_4 and 23.4 slm H_2O; GHSV≅ 2500 h^{-1}. Adapted from [2, 58].

The experimental results are obtained feeding CH_4 and H_2O (S:C = 3) to a uniformly preheated packed bed at the initial temperatures of 600 and 800 °C respectively. The plots on the top show the good agreement between the measured and the simulated temperature evolution along the reactor at specified times. In analogy to figure 2.3, the temperature profiles in the left plot clearly exhibit a dispersive behaviour, given that the initial preheating temperature of the catalyst lays below T_{crit}. On the contrary, if the initial temperature of the bed is higher, a dispersive and a compressive region in the temperature front combine (right). This effect can be clearly observed in the lower plots, which show the measured temperatures over time at some of the positions indicated in the upper figures. The substantial difference between the curves in the left and right plots and, in particular, the rapid temperature decrease as the front travels over the measurement positions in the right plot, corroborate the occurrence of the postulated endothermic steep front.

In further studies, the effect of increasing the temperature of the feed on the characteristics of the evolving front has been analyzed. In such a case, it seems obvious that the temperature of the feed must drop in the reaction zone due to the energy consumption by the endothermic reaction, resulting in a stationary step profile. As reported by Glöckler et al. in [3], the lowest temperature $T_{eq,stat}$ is implicitely defined by the following relation

$$\underbrace{(\dot{m} \cdot c_p)_g(T_{up} - T_{eq,stat})}_{\textit{change in sensible heat}} = \underbrace{\frac{\dot{m}w_0}{MW}(-\Delta h_r)}_{\textit{heat consumption by reaction}} \tag{2.13}$$

$$T_{up} - T_{eq,stat} = |\Delta T_{ad}| \cdot X(T_{eq,stat}) \tag{2.14}$$

being $|\Delta T_{ad}|$ the hypothetical adiabatic temperature drop at complete conversion. Interestingly, if $T_{eq,stat} > T_{low}$ (cf. fig. 2.3 c), no dispersive region evolves. In turn, both a stationary reaction zone at the inlet of the active catalyst, characterized by a temperature drop $\Delta T = T_{up} - T_{eq,stat}$ and a travelling front evolve. The velocity of the front can be estimated, in analogy to equation (2.10), according to following expression

$$w_{shock} = \underbrace{\frac{(\dot{m} \cdot c_p)_g}{(1-\varepsilon) \cdot (\rho \cdot c)_s}}_{w_{therm}} \cdot \left(1 + \left|\Delta T^{ad}_{endo}\right| \cdot \frac{X(T_{up}) - X(T_{eq,stat})}{T_{up} - T_{eq,stat}}\right) \tag{2.15}$$

Experimental results corroborating these behaviour have been also reported and are summarized in figure 2.5, where an inert section is placed prior to the active zone of the packed bed. The feed is preheated in this section and enters the catalytic zone at a temperature equivalent to the preheating temperature of the packed bed.

The relevance of this behaviour for the design of a reverse-flow reactor has been introduced in [3] and is recalled in section 7.2 of this work.

Regeneration period

As already discussed, the aim of the regeneration period is to reestablish the desired initial temperature profile for the production period (s. figure 2.2). Flow reversal and the corresponding convective displacement of the temperature profile in the new flow direction contributes to this purpose. The velocity of the thermal front corresponds to the term w_{therm} in equation (2.10). However, the energy consumed during the production period has to be additionally supplied. In [1, 39], Glöckler described an analysis of the optimal strategy to introduce energy into the system. The procedure consisted in determining in a recursive

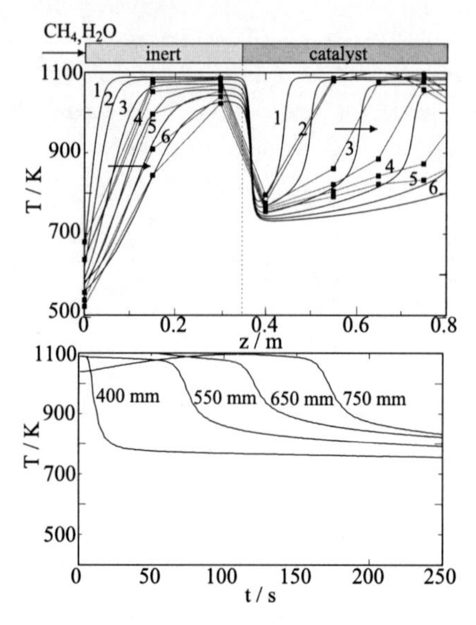

Figure 2.5: Temperature evolution during methane steam reforming in a uniformly preheated packed bed, consisting of an inert section followed by an active section. Top: measured (points) and simulated (straight lines) temperatures along the packed bed at 25, 50, 100, 150, 200 and 250 s after reaction beginning (profiles 1 to 6 respectively). Bottom: temperature evolution over time at the specified positions. Feed: 7.8 slm CH_4 and 23.4 slm H_2O; GHSV$\cong$ 2500 h^{-1}. Initial preheating temperature $\cong$ 800 °C. Adapted from [2, 3, 58, 59].

manner a constant, axially distributed heat source profile ($\dot{q}_{exo}(z)$), which enables the establishment of the desired initial temperature profile after the regeneration period. Making use of the convective energy transport after flow reversal, this can be materialized by introducing several discrete heat sources separated by the distance travelled by the thermal front during the regeneration period z_{shift}, as described in [1] and summarized in figure 2.6. The suitability of this regeneration concept has been validated based on dynamic simulations with a model accounting for dispersive effects.

In practice, energy is released at each of the distributed heat inputs by combustion of a fuel gas. Since oxidizing conditions in the catalytic section are to be avoided at high temperatures, the obvious procedure consists of feeding the fuel together with the main flow and injecting oxygen at several positions along the reactor length. The proper positioning of the air nozzles and the regeneration concept itself are closely related to the characteristics of the temperature profile evolving during the production period.

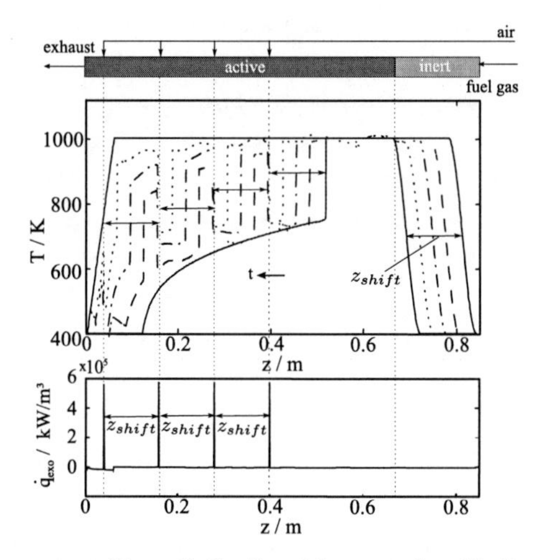

Figure 2.6: Temperature profiles and distributed heat supply with discrete energy point-sources along the reactor length during the regeneration period based on a simplified model. Adapted from [1].

It is worth noticing that, in analogy to equation (2.10), the velocity of a moving exothermic reaction front is defined as

$$w_R = \underbrace{\frac{(\dot{m} \cdot c_p)_g}{(1-\varepsilon) \cdot (\rho \cdot c)_s}}_{w_{therm}} \cdot \left(1 - \frac{\Delta T_{exo}^{ad}}{\Delta T_R}\right) \tag{2.16}$$

being ΔT_R the effective temperature rise of the exothermic front. During the regeneration period, the reaction front should stabilize at each of the discrete side feeds, whereas the packed bed is reheated by the downstream energy transport occurring with the velocity of a pure thermal wave. Therefore, w_R must tend to zero, which can be achieved whenever the temperature of the reacting mixture entering the combustion zone lays above its ignition temperature.

The basic features of the production and the regeneration steps can be condensed into a short cut method for the design of the RFR as discussed in the following section.

2.3.2 Short cut method for reactor design

The key feature of the RFR concept is the formation of the self sharpening, endothermic temperature front that enables an efficient usage of the energy stored in the packed bed if previously preheated beyond a threshold temperature (cf. fig. 2.3). This reaction front travels in flow direction with a larger velocity than that of a pure thermal wave.

The velocity of the compressive or *shock* region within the endothermic temperature front w_{shock} can be estimated combining equations (2.10) and (2.12). The resulting relation between the reaction front and the thermal wave velocities w_{shock}/w_{therm} is shown in figure 2.7 in agreement to data published in [3].

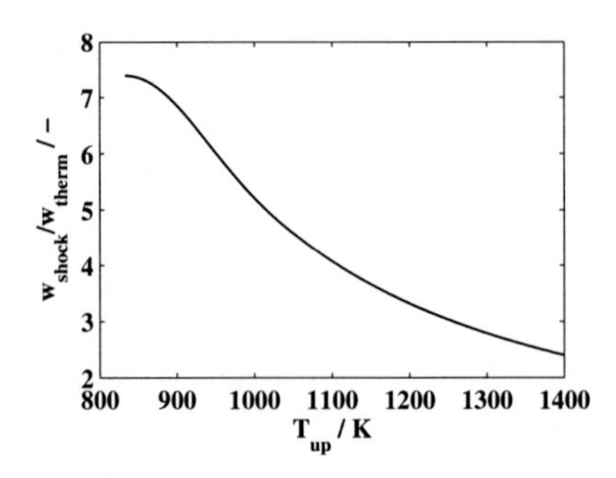

Figure 2.7: Shock and thermal wave velocities relation in a catalytic packed bed depending on the preheating temperature T_{up}.

Setting maximal conversion during the endothermic period as a constraint, the maximal duration of the production step is given by

$$\tau_{prod} \leq \frac{L_{active}}{w_{shock}} \tag{2.17}$$

with L_{active} accounting for the reactor length filled with active packing. Accordingly, the length of the inert packing L_{inert} of the reactor (cf. fig. 2.2) should account at least for the distance z_{therm}, travelled by the thermal front during τ_{prod}

$$z_{therm} = w_{therm} \cdot \tau_{prod} \tag{2.18}$$

$$L_{inert} \geq z_{therm} \tag{2.19}$$

The effective length of the inert section has a direct effect on the maximal temperature increase of the reactor effluents $\Delta T_{effluent}$. This increase can be estimated based on the evaluation of the temperature slope in the inert heat exchange section. This can be obtained from an energy balance at the reactor boundary, under the assumption that the temperature slope remains constant over the length of the heat exchange section. Accordingly, the expression resulting for the inert section at the outlet of the RFR extracted from [39] reads as follows

$$\Delta T_{effluent} = \frac{T_{up} - T_{feed}}{2 + \frac{Pe}{d_{hydr}} \cdot (L_{inert} - w_{therm} \cdot \tau_{prod})} \tag{2.20}$$

Based on the above discussed considerations and defining the maximal allowed temperature of the reactor effluents during production, the reactor can be dimensioned and the core operation parameters (mass flow density $\dot{m}$ and production time τ_{prod}) estimated.

During the regeneration period, the desired temperature profile to run the endothermic semicycle has to be established. As a first approach, the duration of the regeneration period τ_{reg} equals that of the production one, in order to enable simultaneous operation of two alternating reactors. For a given reactor design, the choice of the operation parameters must fulfill the following conditions:

- the integral energy balance over the two periods must be weakly exothermic,
- the reaction zones must remain within the packed bed (zero differential creep velocity condition [4, 37, 60]), and
- equal heat capacity fluxes during both periods should be striven for, in order to optimize heat recovery.

If the first condition is not fulfilled, the temperature level in the active zone will successively decrease until the reactor is extinguished. On the contrary, if the system is overheated, largest productivity can be achieved but energy integration is insufficient. The short cut method to determine the regeneration conditions originally focused on the first condition, assuming equality of heat fluxes during production and regeneration and thus, fulfillment of the second and third conditions. Thus, a complete analysis of the system behaviour can be performed on a theoretical basis by defining the position of the energy sources according to equation (2.18) and adjusting the energy input in order to fulfill the first of the conditions listed above (cf. fig. 2.6).

It should be noticed that the short cut method relies on a simplified picture of the system that neglects variations in the heat fluxes along the reactor length and over a complete operation cycle. However, the heat flux is affected by mass flow variations and the state dependency of the physical properties of the flow. As already introduced by Bernnat and Glöckler in [61], both aspects can lead to noteworthy deviations from the simplifying assumption in a real system. The effects of these deviations on the real behaviour of the RFR are analyzed in the following chapters.

2.3.3 Experimental validation and technical limitations

An experimental proof-of-concept of the RFR was performed and reported by Glöckler et al. in [11, 12, 40], based on a reactor design resembling that sketched in figures 2.2 and 2.6, i.e. the packed bed consisting of an active zone and an inert section located at the reactor outlet during the production period. The experimental results, however, did not enable to corroborate an unambiguous formation of the endothermic shock wave previously described. Moreover, the inspection of the reactor after the experimental tests revealed the occurrence of excess temperatures in the position of the air injectors, resulting in irreversible damages in the construction and the catalyst around the nozzles as shown in figure 2.8. Detailed simulations of the system were performed in order to get a better insight into the origin of these observations [11, 61].

Whereas the first aspect could be partially attributed to the dispersive effects of the steel elements in the reactor construction (reactor wall and air conductions) on the temperature profile evolution, as well as on a lower catalyst activity than expected, the problems regarding the occurrence of excess temperatures during regeneration seemed to have its origin in the ignition of homogeneous combustion reactions around the air nozzles.

Figure 2.8: Thermally damaged and deformed air injection nozzles after operation (left and middle) and sintered catalyst particles (right).

This finding suggested that avoiding catalytic packing around the injectors could contribute reducing the energy released by heterogeneous combustion around the air injectors and thus, suppress the ignition of homogeneous reactions. Furthermore, the use of fuels with elevated homogeneous ignition temperatures (e.g. CH_4) or addition of inhibiting inert substances to the regeneration gas mixture (e.g. H_2O, CO_2) could further contribute minimizing the occurrence of excess temperatures. In a further set of experimental results reported in [41], the packing around the air nozzles was indeed replaced by sections of inert packing. Moreover, the air supply during the regeneration period was subject to a feed-forward control strategy, which reduced the air flow as the regeneration proceeded and the temperature level in the system increased. However, appearance of excess temperatures and material damage in the combustion sections could not be fully avoided. Measured temperature profiles in periodic operation under such conditions are shown in figure 2.9, where it can be seen, that the thermocouples in the combustion section recorded temperatures close to 1200 °C.

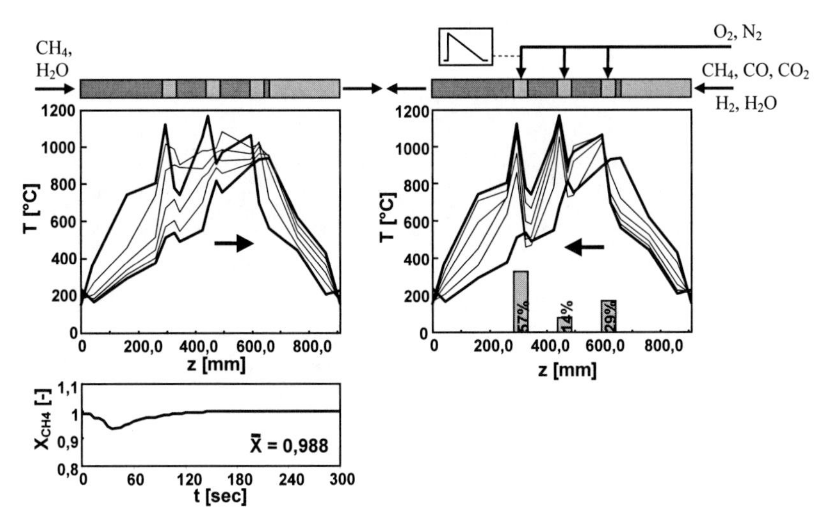

Figure 2.9: Measured temperature profiles and methane conversion during the production (left) and the regeneration period (right) under periodic steady state. Each temperature profile corresponds to time intervals of 60 s. Feed composition during production: $S:C = 3:1$; 7 slm CH_4. Extracted from [41].

Several issues were identified as weak points in the strategy pursued. The energy stored in the inert packing around the air nozzles was transported convectively during the production period, whereas no significant cooling effect through the endothermic reaction occurred. Thus, a high temperature level remained during production, which favoured the homogeneous ignition of the fuel mixture during the subsequent regeneration. The effect of inhibitors (e.g. H_2O, CO_2) was negligible and the exclusive use of methane as fuel to increase the homogeneous ignition temperature was coupled with its thermal decomposition downstream and accordingly, with catalyst deactivation by coke deposition. In turn, use of steam to supress coking caused strong cooling of the packed bed during regeneration through reforming of remaining methane, causing a significant loss in productivity [62].

These results represent the starting point of the current analysis, which aim at the development of an alternative regeneration strategy for the described RFR.

2.4 Summary

The potential of the RFR operation for hydrogen synthesis based on MSR has been highlighted in the previous sections. Overcoming the heat transfer limitations of most of the conventional technologies is the most relevant attribute of this concept. Its validity has been summarized based on theoretical and experimental studies and the limitations in the realisation of the concept have been identified taking previous research work as reference. These limitations are closely related to the regeneration step and the efficient reheating of the packed bed at high temperatures, without exceeding levels at which the construction materials or the reactor packing could be damaged.

Reconsidering the way to introduce energy into the system in a controlled manner represents a crucial aspect in order to overcome the technical hurdles challenging the RFR concept. Accordingly, this aspect turns into a core element of the current study and its analysis is reported in the following chapters.

Chapter 3

Flameless combustion for heat supply in a packed bed

A crucial aspect for the successful operation of the RFR is the efficient reheating of the packed bed at a high temperature level. As previously discussed, temperature maxima occurring during combustion might damage the reactor internals and deactivate the catalytic packing. The adaption of the flameless oxidation (FLOX®) concept represents an interesting alternative to introduce the required heat directly into the packed bed, avoiding excess temperatures and overcoming unnecessary heat transport limitations. The operation principle relies on the suppression of flame formation during combustion and thus, on the limitation of the maximal temperatures. In such a case, the temperature rise in the combustion zones approaches the adiabatic temperature increase of the combusting mixture under ideal adiabatic conditions. Hence, if FLOX® conditions are established, reheating of the system at the desired temperature level can be achieved by control of the fuel and air being fed to the combustion chambers.

3.1 Flameless Oxidation (FLOX) and its conventional applications

The concept of flameless combustion is well known and already successfully implemented in stationary applications under fuel lean (i.e. excess of oxygen) conditions. The classical example of FLOX® burner application is the steel industry, whereas new fields are continuously being explored such as integrated MSR systems (cf. section 2.2), clean combustion of lean gases deriving e.g. from agricultural activities, firing of pressurized pulverized coal in coal-fired power plants or glass production and processing [63–66].

This form of combustion finds its origin in an attempt to reduce the NO_x emissions in industrial burners, in order to fulfill the stringent environmental regulations. This is of special interest for self-recuperative burners. The latter preheat the combustion air using the energy contained in the combustion off-gases in order to reduce energetic and operational costs. However, the increase of the maximal temperatures reached in the burner directly affects the amount of thermal NO_x formed [42–44]. The introduction of secondary measures to reduce the emissions of nitrogen oxides via after-treatment of the combustion off-gases is costly and cannot be taken into consideration for small facilities. Thus, advanced combustion techniques such as staged-air low NO_x burners, ultra-low NO_x burners and in particular FLOX® burners can contribute effectively reducing the emissions even at high air preheating rates [67, 68].

The FLOX® operation principle is based on the mixing of a large amount of combustion flue gases with combustion air before the latter reacts with the fuel [42]. A proper homogeneous mixture of these three components (air, fuel and off-gases) suppresses the formation of defined reaction fronts (flames), which normally account for adiabatic flame temperatures around 2000 °C for most of the hydrocarbons used as fuel. The flame temperature can reach even higher values depending on the air preheating rate.

In a combustor operating under FLOX® or also called mild combustion conditions, recirculation of flue gases increases the inert amount contained in the mixture. Thus, not only the reaction front is dispersed due to the strong turbulence generated in the chamber, but the adiabatic temperature increase in the reaction zone is also attenuated in agreement to the operation principle of flue gas recirculation as a technique to control NO_x emissions [67]. Consequently, the recirculation rate K_V of the combustion off-gases turns into a decisive parameter for the operation of a burner. The recirculation rate is defined as

$$K_V = \frac{\dot{M}_E}{\dot{M}_F + \dot{M}_A} \tag{3.1}$$

where $\dot{M}_E$ describes the flow of recirculated exhaust gases in relation to the sum of fuel, $\dot{M}_F$, and air flows $\dot{M}_A$. Recirculation of exhaust gases as a NO_x-reduction measure requires rates of $K_V \leq 0.3$, whereas higher recirculation rates must be avoided in order to ensure the stability of the flame. However, according to Wünning et. al [43, 44], stable combustion is also possible at much higher recirculation rates if sufficiently high temperatures in the burner can be maintained. A peculiarity of the operation for large K_V-values is that combustion occurs in absence of visible or audible flames. Thus, depending on the combination of both parameters, several combustion regimes may occur, with recirculation rates between 2 and 3 being sufficient to ensure stable flameless combustion (see fig. 3.1).

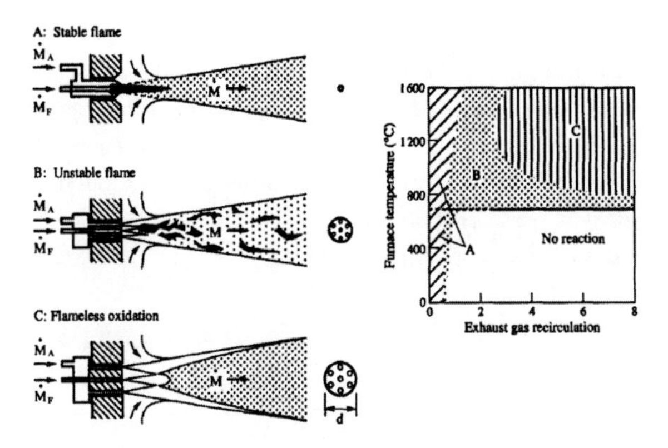

Figure 3.1: Combustion regimes with stable (A) and unstable flames (B), and under flameless combustion (C) as a function of the recirculation ratio. Extracted from [44].

Based on the last considerations, temperature and recirculation rate of the exhaust gases are the decisive parameters to ensure FLOX® conditions in the combustion chamber. In order to enter the FLOX® region, the combustion chamber must be heated up in flame mode and then switched to flameless operation. This can be achieved by performing a switch in the air injection nozzles. Under FLOX® conditions, an appropriate geometrical arrangement of such nozzles enables the proper mixing of the air jets with exhaust gases before getting in contact with the fuel. The effect of the jet velocity amongst other operation parameters, such as the nature of the fuel used, has been reported by Derudi et al. [69, 70].

Despite the great attention that mild combustion technology has been and is currently being paid, introduction of this concept into a packed bed and in particular into a periodically operated RFR for reheating purposes, poses several challenges that are addressed in detail in the following section.

3.2 Flameless combustion concept in a packed-bed reactor

The development of a combustion chamber to be operated within a packed bed implies remarkable differences not only in the geometrical design but also in the operation, compared to that of a FLOX® burner for conventional applications. The combustion chamber must be embedded in the packed bed, thus structuring the latter in a sequence of empty (combustion chamber) and packed sections. The latter can be filled either with catalytic or inert packing depending on the configuration and structure of the different zones in the reactor. In direct analogy with the latest experimental results reported by Glöckler et al. in [41], the combustion chamber substitutes the inert zone around the air nozzles and thus, is located between two zones filled with active packing. It is worth noticing that the chamber itself represents a void section within the packing. Accordingly, the heat capacity of the combustion zone can be significantly reduced in comparison to previous reheating concepts and the problem of undesired heat storage in the combustion sections reported in previous studies can be successfully minimized (cf. section 2.3.3).

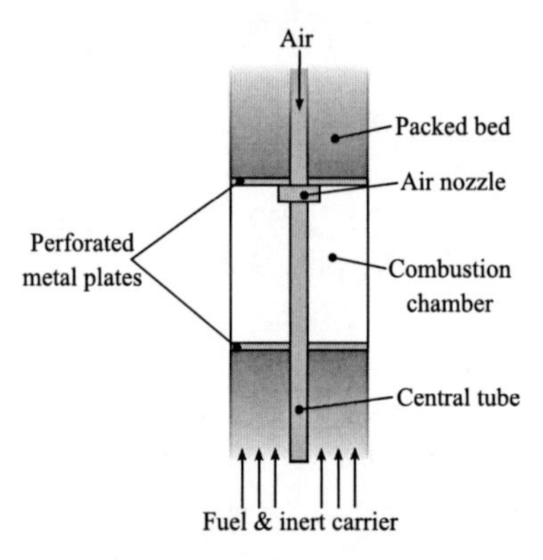

Figure 3.2: Representative configuration of a flameless combustion chamber embedded in a packed bed.

The combustion chamber is conceived to allow the introduction of a fuel mixture, an inert flow and the air required for combustion. The chamber can be conceptually represented by a similar construction to that depicted in figure 3.2. It is delimited by two perforated metal plates that separate the combustion from the packed zones, serving as a structuring element for the packed bed. The diameter of the chamber equals the reactor diameter and the height can be modified by adjustment of the distance between the upper and lower plates. These are fixed at a central tube that stabilizes the reactor internals and supplies the chambers with combustion air (side stream). The central tube is surrounded by packing over almost its whole length. After the reforming semicycle, the packing has an average temperature of 500 to 600 °C. Thus, the energy stored in the packing is partially used to preheat the combustion air before entering the combustion chambers. The fuel together with an inert carrier is fed to the bottom of the reactor and enters the combustion chamber through the lower plate.

Ideally, an ignitable mixture is formed and combustion occurs homogeneously and under flameless conditions in the void of the chamber. Therefor, the chamber temperature must exceed the homogeneous ignition temperature of the gas mixture. If this is not the case, the mixture reacts heterogeneously in the catalytic active section at the outlet boundary of the chamber, where temperature rapidly increases. Ideally, this high temperature region heats up the chamber volume and assumes the function of an ignition source, turning the heterogeneous combustion regime into a homogeneous one. Several aspects that might influence the ignition behaviour in the chamber, as well as experimental results providing a better insight of this process under experimental conditions, are reported in section 3.4.2.

The different characteristics of the combustion chamber and its operation principle in comparison to conventional FLOX® burners are summarized below:

1. The generation of a strong recirculation of the gases in the chamber is decisive to operate in FLOX® regime. In conventional FLOX® burners, fuel and air injectors are geometrically designed in order to generate a sufficient recirculation of the exhaust gases. In contrast, the current concept for the combustion chamber only foresees nozzles for the injection of combustion air. The latter must be appropriately designed in order to ensure a sufficient recirculation of the gases in the chamber.

2. Contrary to conventional FLOX® burners, which can be run at several operation modes (e.g. start up in flame mode, flameless operation), the combustion chamber will be operated in one mode during the whole regeneration period. The presence of catalytic packing at the inlet and outlet of the combustion chamber poses a major difference to conventional burners. Its influence in the ignition behaviour of the gas mixture in the chamber needs to be analyzed.

3. Conventional burners are generally designed to operate under air rich conditions to ensure the total oxidation of the fuel and to partly reduce the adiabatic temperature increase. Based on the current reactor concept, stable operation under flameless conditions is required for fuel rich conditions ($\lambda < 1$)[1], given that the total amount of fuel is fed together with an inert gas carrier at the inlet of the reactor. The heat release at every side port is therefore controlled by the amount of combustion air added. The stability of flameless combustion under such conditions at a reheating temperature of 1000 °C must be validated.

[1] λ defines the ratio between the introduced air/oxygen for combustion and the stoichiometric amount required for the total oxidation of the fuel.

4. The periodicity of the process implies transient inlet temperatures of the main flow entering the combustion chamber. Initially, the inlet temperature is around 500 to 600 °C and increases (in the second and further chambers) or slightly decreases (first chamber) during the reheating period.

5. In reference to the two previous points, a feed-back control strategy for the combustion chamber that enables reheating at a constant temperature despite the transient behaviour of the process parameters has to be defined and validated.

The suitability of the FLOX® concept for *in situ* reheating of a packed bed has been tested in a row of preliminary experimental and simulation studies that are briefly summarized in section 3.3. The insight obtained during the preliminary work sets the basis for a systematic development of an appropriate combustion chamber geometry and operation procedure, taking the above mentioned aspects into account and making use of several theoretical and experimental methods that are described in the same section. The operation characteristics of a chamber developed on such a basis has been analyzed experimentally as summarized and discussed in section 3.4.

3.3 Suitability analysis and combustion chamber development

The combination of exploratory experimental work and standard computational fluid dynamics delivered first information about the potential of the FLOX® concept to reheat a packed bed by means of *in situ* combustion. As already mentioned in the previous sections, one of the crucial aspects is to achieve large recirculation rates within a combustion zone of reduced dimensions. The reference geometry consists of a cylindrical chamber of 50 mm inner diameter and 50 mm height, representing about 5% of the total volume of the reactor prototype used in the experimental setup (cf. section 5.1). Given this constraints, the behaviour in the chamber itself is determined by the flow pattern during combustion. The latter is, in turn, controlled by the well-directed injection of reactants and more specifically, of combustion air. This configuration takes the original regeneration concept of the RFR as reference. The main flow, consisting of a fuel blend diluted in an inert component, is fed through one reactor end and enters the chamber through the lowest plate (see figures 3.2 and 3.3). Combustion air is injected from the upper boundary in opposite direction to the main flow. The exact position and geometry of the injection nozzle is decisive for the formation of the required recirculation pattern to achieve flameless operation.

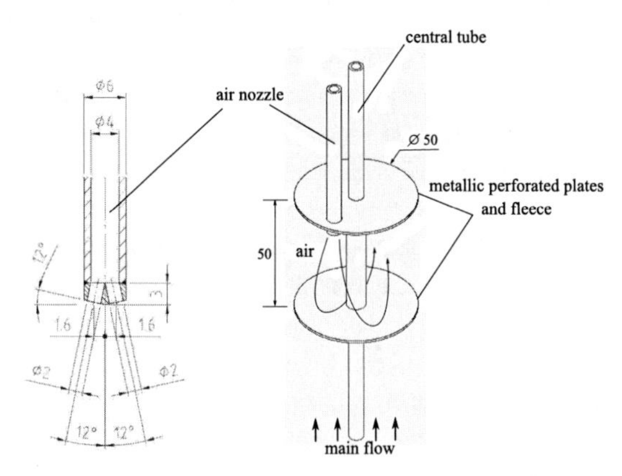

Figure 3.3: Original geometry for exploratory studies of a flameless combustion chamber embeded in a packed bed. Extracted and adapted from [71].

The original construction tested within the frame of this work is described by Rink [71] and its operation is based on a relatively simple flow configuration that uses the vertical symmetry plane of the chamber to virtually divide it into two half-cylinders. The air injector has two circular nozzles, responsible for the recirculation in each of the two sections respectively (see figure 3.3). By displacing the injector from the center of the chamber toward an outer radial position, a self induced recirculation in the vertical plane can be achieved.

The experimental observations proved the applicability of the concept and delivered interesting information to be taken into account for a subsequent design of the combustion chamber. In order to do so, the combustion chamber and its behaviour was reproduced by means of CFD simulations, as described in the following.

3.3.1 CFD-Simulations

The behavior observed experimentally was qualitatively reproduced by means of CFD simulations [72]. These were mostly run under stationary isothermal conditions. A few simulations under non-isothermal conditions were performed for comparison by coupling the calculation of the flow field with a reduced reaction mechanism for natural gas combustion based on the GRI mechanism [73]. Despite the differences in the flow pattern introduced by the changes of the physical properties in the reacting flow, the performance indicators used to describe the recirculation ratio and the mixing quality in the combustion chamber exhibit similar trends for both the isothermal and non-isothermal cases. Thus, non-reacting isothermal calculations are primarily used, since they yield an adequate description of the flow pattern in the chamber and are computationally less expensive. Monz summarizes in [72] some of the most relevant aspects to be taken into account when conceiving an appropriate chamber geometry. Due to the reduced dimensions available, taking advantage of the swirl effect of a helical flow pattern as conventionally used for flame stabilisation [74] improves homogeneity, minimizes segregated zones and contributes avoiding flow penetration into the neighboring packing, even if the injected flow exhibits a large momentum. The latter turned out to be a main issue in preliminary tests of the combustion chamber, leading to the formation of hotspots in the packing below the chamber [71].

On the basis of a further simulation study varying the number, position and setting angle of the injection nozzles, the geometry sketched in figure 3.4 was identified as most favorable. The flows used for the design correspond to the fuel, air and inert flows required during regeneration of a 1 m long reactor, which produces the amount of hydrogen corresponding to a continuous thermal power of $5\,\mathrm{kW}_{\mathrm{LHV,H_2}}$[2] during reforming. This operation regime has been chosen as reference in analogy to the latest RFR experimental results reported by Glöckler et al. in [41].

The parameter estimation is made based on following consideration: the reaction energy Q_{prod} consumed during the production period must be regenerated by combustion of PSA off-gas. Assuming conservative heat transfer efficiencies of only about 70% during regeneration, the total amount of fuel and air required for combustion can be estimated. Since the reactor in [41] has three combustion zones, the flow ranges in each chamber result from the combustion of 20 to 35% of the introduced fuel. This proportion can be directly related with λ, which indeed has been varied in a wider range ($0.2 \leq \lambda \leq 0.6$). Furthermore, an inert carrier must be added in order to balance the heat flux and consequently, the thermal front velocities between the production and regeneration periods (s. eq. 2.10). Different flow configurations for several reforming loads are summarized in table A.1.

[2]LHV, Low Heating Value

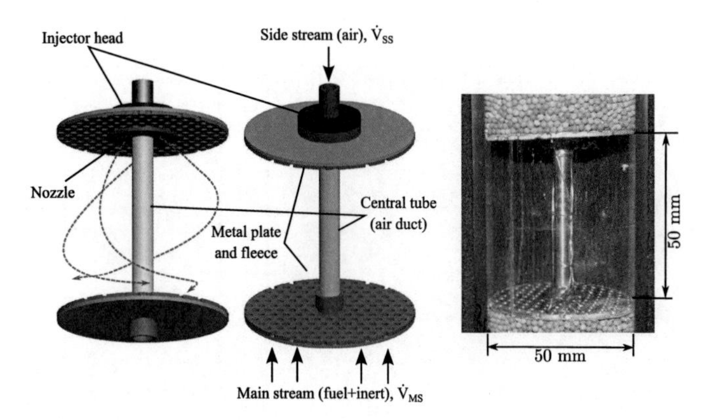

Figure 3.4: Combustion chamber geometry. Left: 3-D representation; right: chamber placed between catalytic packing in a quartz glass reactor for combustion experiments.

In the combustion chamber depicted in figure 3.4, air is injected counter-currently to the main flow through a radially centered nozzle head located at the top of the chamber. Several drill-holes tangentially oriented to the outer boundary of the lower chamber plate are used to inject the air with a rotational impulse. In conjunction with the flow entering the chamber through the lower plate, a toroidal eddy is generated enhancing the recirculation of the gases. Besides this, proper tuning of the pressure drop generated by the perforated plates is essential to hinder the penetration of the injected air into the catalyst packing located upstream.

Recirculation of the gases as well as the degree of mixing are parameters on which the chamber design and optimization is based (see section 4.3.1). As shown in figure 3.5, the maximum simulated recirculation ratio of the present configuration clearly indicates that for large air volume flows (i.e. $\lambda \approx 0.6$)[3] values of $K_V > 3$ can be achieved. This is above the minimal recirculation ratio of 2 to 3 that should suffice to attain flameless combustion regimes (cf. fig. 3.1). For lower combustion loads ($\lambda \leq 0.2$) maximum recirculation ratios of 1.43 are calculated. In this case, the momentum of the injected air does not ensure a complete suppression of flames as will be further experimentally corroborated (cf. sec. 3.4.1).

[3] The simulated recirculation ratio for $\lambda = 0.6$ has been generated with a parameter set fulfilling the condition $\frac{\dot{V}_{SS=5.64slm}}{\dot{V}_{MS}} = \frac{\dot{V}_{SS,\lambda=0.6}}{\dot{V}_{MS=50.5slm}}$, i.e. the volume flows represent 20% of the flows reported in table A.1 for a reforming load of 5 kW_{LHV,H_2}. As discussed later on, K_V is more sensitive to the side-to-main volume flow ratio than to the volume flow in absolute terms.

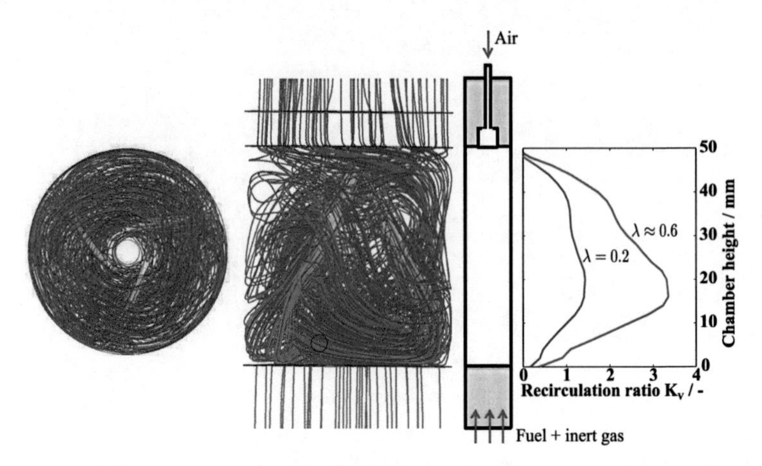

Figure 3.5: Representative results of the preliminary simulation study to identify appropriate chamber geometries. Left: horizontal and vertical view of the streamlines computed with CFD; right: recirculation ratio over the chamber height for different operation regimes during regeneration of a RFR operating at a load of $5\,kW_{LHV,H_2}$ (cf. tab. A.1).

3.3.2 Tracer Methods

CFD simulations are a powerful tool for a preliminary conception and performance prediction of the combustion chamber. However, a systematic optimization task including parametrization and geometry improvement implies a considerable computational effort. Furthermore, the information delivered by the stationary simulations cannot be extrapolated to describe the dynamic performance of the system. This motivates the utilization of tracer methods, in order to give a solid answer to the first question arising in section 3.2. An exact reproduction of the combustion chamber has been therefor tested in a water channel. For visual inspection, a Plexiglas mock-up with analogous geometry and dimensions (inner diameter 50 mm) was built. The internals of the chamber are made of stainless steel and the packing is modeled using 3 mm glass pellets. The construction allows a relatively simple exchange of the air injection nozzles, so that several geometries can be efficiently tested over a wide range of operation conditions.

The operating conditions for the water flows are defined making use of the Reynold's similarity of the flows. Since geometric analogy is given, process similarity can be guaranteed if the dimensionless numbers describing the process are numerically equal [75]. This consideration allows a further extrapolation of the behavior observed in the water mock-up to the behavior under real conditions in the gas phase. As most important dimensionless number, Re is defined as follows:

$$Re = \frac{\rho \cdot v \cdot d_h}{\mu} \tag{3.2}$$

The adjustment of the parameters to resemble the jet flow at the air injector nozzles is made considering the jet velocity and setting d_h to the diameter of one nozzle. In turn, the adjustment of the parameters to resemble the characteristics of the main stream is made according to the following definition:

$$Re = \frac{Re_p}{(1-\varepsilon)} = \frac{\rho \cdot v \cdot d_p}{\mu} \cdot \frac{1}{(1-\varepsilon)} \tag{3.3}$$

The Reynolds number based on the so called particle Reynolds number Re_p and the void fraction of the packing ε describes the flow through the packed bed. In this case, the velocity v represents the superficial velocity of the flow, whereas d_p is the particle diameter of the packing.

The adjustment of the water flows to equal gas operation is made based upon the physical properties at a temperature of 25 °C for water and 475 °C and 1 bar for gas respectively.

Real-time recording

Tracer experiments were recorded both in real-time and with help of a high-speed camera. For this purpose, blue ink is continuously added as tracer with help of a peristaltic pump to the water flow that enters the chamber through the injector nozzle. In this way it is possible to detect if dead zones or segregation occurs and thus, identify in an early stage those constructions that are not suitable for the purpose pursued. Moreover, addition of small amounts of ink through a small cannula located at the bottom of the chamber allows an assessment of the trajectory of the stream lines within the chamber.

Based on the real-time recordings is it possible to efficiently test the mixing performance of the combustion chamber in a wide range of operation parameters. The latter are partially summarized in table A.3 and should serve as an outline of the order of magnitude of the flows tested.

The current configuration of the chamber and the several injectors tested exhibit good homogeneity characteristics, especially for large operation loads and/or side-to-main volume flow ratios $\frac{\dot{V}_{SS}}{\dot{V}_{MS}}$, which represents a measure of the flow being injected through the nozzle and the flow entering the chamber through the bottom plate. Hence, in order to highlight the differences between the various nozzles tested, the experimental results shown in figure 3.6 correspond to a low side-to-main volume ratio (equivalent to combustion with $\lambda = 0.1$). Blue ink is added as tracer to the water flow injected through the air nozzles. The main flow entering the chamber from below mixes with the dyed flow and leaves the chamber through the upper boundary. In continuous operation, differences in the color concentration serve to visually assess the degree of mixing in the chamber (figure 3.6, top). Dark zones correspond to a higher tracer concentration and are an indicator of poor mixing. Below each capture, its respective analogous with inverted colours for easier interpretation is also shown.

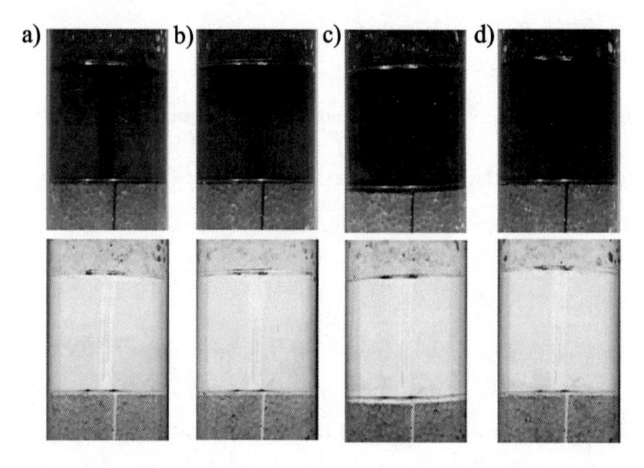

Figure 3.6: Experiments for visualization of the degree of mixing with continuous addition of tracer. Real captures (up) and with inverted colours (bottom) for a main flow of $1.02\,\frac{\mathrm{l}}{\mathrm{min}}$ (3.36 kW, cf. table A.1) and a side flow of $0.16\,\frac{\mathrm{l}}{\mathrm{min}}$ ($\lambda = 0.1$) respectively. Variation of the number (N), the angle (A) and the diameter of the nozzles (D). Adapted from [76].
a) N = 3; A = 47°; D = 1.0 mm
b) N = 4; A = 47°; D = 1.0 mm
c) N = 3; A = 64°; D = 1.5 mm
d) N = 4; A = 64°; D = 1.0 mm

Comparison of the different pictures leads to the conclusion that the lower zone of the chamber is better mixed for those configurations with a 3-nozzle injector than for those with a 4-nozzle one. This effect is due to the increased momentum of the injected flow in the former case and correlates with the general observation that, for a given geometry, homogeneity improves for large side-to-main volume flow ratios $\frac{\dot{V}_{SS}}{\dot{V}_{MS}}$. Besides this, the tracer distribution within the whole chamber volume is again improved when the side flow jet creates an angle of 64.7° with the horizontal plane, as the flow injected is directed to the junction between the chamber wall and the lowest metal plate, maximizing the distance between the nozzle exit and the lower boundary of the chamber. This configuration has two positive effects. The length of the jet and hence, the relative velocity between the latter and the surrounding fluid is maximized, contributing to a larger entrainment of bulk fluid and improving homogeneization [77–79]. Furthermore, the momentum at the lowest position of the combustion chamber can be reduced. Hence, penetration of the injected flow in the lowest packing could not be observed for any of the configurations and $\frac{\dot{V}_{SS}}{\dot{V}_{MS}}$ rates tested.

High-speed camera recording

Besides the analysis described above, captions made with a high-speed camera allow a closer analysis of the mixing process dynamics. It is thereby important to obtain a reliable estimation of the mixing time required to achieve sufficient homogeneity in the chamber. It must be noted, that mixing time in the current case refers to macro-mixing and it is assumed to be the rate limiting process during combustion. It is generally accepted that average residence times of at least one order of magnitude larger than the mixing times are necessary in order to achieve homogeneous mixing in a tank [80]. In order to determine the mixing time, the methodology followed was inspired by the analysis method described by W. Müller in [81]. The latter uses a very well defined light source that illuminates a cylindrical glass model of a stirred tank from behind. In order to determine the mixing homogeneity, a tinted flow is intermittently injected. Since the extinction of the light beam correlates with the concentration of the tracer, the degree of mixing in the container can be derived from the statistical analysis of the gray tone distribution of a set of photographic captures of the system [81, 82].

Contrary to the experimental setup and procedure described by Müller, which pursued an accurate determination of the degree of mixing in the system, main focus of the present analysis is to obtain a qualitative estimation for the required mixing time. The experimental setup used in the water mock-up is subject to several interferences and uncertainties such as the quality of the light beam used, the reproducibility of the tracer concentration or even the quality of the captures. Thus, a direct correlation between the gray tone distribution in the captures and the degree of mixing in the chamber is omitted. Nevertheless, the evaluation of the results is based in a similar procedure to the one described above. Figure 3.7 depicts schematically the proceeding followed to analyze the results.

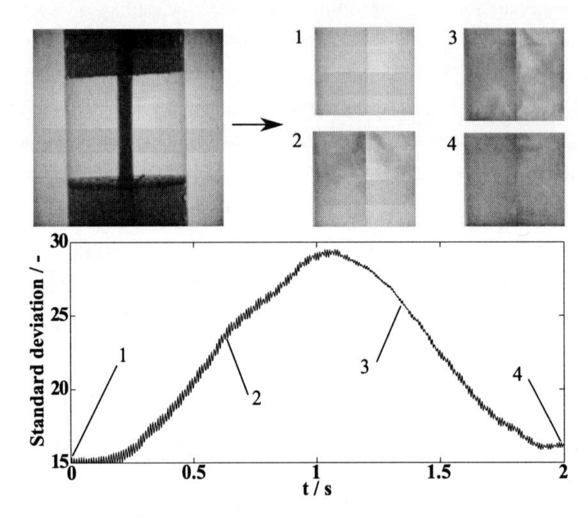

Figure 3.7: Schematic representation of the procedure to analyze the captures taken with the high-speed camera. Top, left: capture of an uncolored chamber; bottom: standard deviation of the gray scale distribution of the pixels over time after injection of the tracer; top, right: analyzed section of the captures at four different times.

The top, leftmost representation corresponds to a capture of the combustion chamber at time $t = 0\,\mathrm{s}$, prior to the injection of a tracer. Since it is illuminated from behind, the packing and the metallic structure appear as dark, opaque zones. The exact position of the latter is defined in terms of the position of the dark pixels, which are not considered in the analysis of further images of a same capture. As soon as the tracer is injected through the air nozzles, the high-speed camera records the dynamic behavior in the chamber with a frequency of 1091 frames per second during a total of approximately 2 seconds and a resolution of 512×512 pixels. During the post-processing, each of the frames is analyzed and the opaque section of the chamber, previously defined, is suppressed (see captures 1 to 4 in the upper, right section of figure 3.7). As already mentioned, for well defined experimental conditions, the variance or the standard deviation of the gray scale distribution of the pixels analyzed can be used as an indicator of the degree of mixing. Large standard deviations correspond to inhomogeneous mixtures and on the contrary, low deviations ($\sigma \approx 0.02$) correspond to almost perfectly homogeneous ones [81]. In the present case, no conclusive correlation between the standard deviation and the degree of mixing can be made. However, setting the deviation at time $t = 0\,\mathrm{s}$ (σ_0) as offset of the measurement and following its evolution over time $\sigma(t)$, it is possible to determine how long it takes, until the mixture in the chamber exhibits such a homogeneity that $\sigma(t_{mix}) \approx \sigma_0$, being t_{mix} the mixing time required for a given set of operation conditions. Based on this interpretation it can be stated that homogeneous mixing is achieved at t_{mix} as soon as $\sigma(t_{mix}) \leq \sigma_0$. The degree of mixing remains invariable for $t \geq t_{mix}$.

Figure 3.7 shows a specific case, in which homogeneous mixing is not fully achieved after 2 seconds. However, the evolution of the mixing process and its correlation with the standard deviation over time can be clearly identified. The configuration corresponds to an injector with three 1.5 mm nozzles with a setting angle of 64.7° and the experimental conditions correspond to a low operation load ($\lambda = 0.1$) and $\frac{\dot{V}_{SS}}{V_{MS}} \approx 0.3$. Provided that larger jet velocities correlate with improved mixing behaviour, all successive analysis and results reported are based on a three 1.0 mm nozzle injector.

The dependency of the mixing time on the different operation parameters is summarized in figure 3.8. Figure a reveals a decaying trend of the mixing time with increasing jet velocity, represented in terms of the Reynonds number in each jet. This behaviour is in full agreement with the general observations in jet mixing phenomena [77–79]. However, even for the largest t_{mix} measured, the residence-to-mixing time ratios $\frac{\tau}{t_{mix}}$ are not far below the value of 10, which represents the generally accepted ratio to ensure a homogeneous mixture. This can be observed in plot b, where the data are plotted against the side-to-main flow ratios $\frac{\dot{V}_{SS}}{V_{MS}}$. The correlation between the latter and Re_{jet} for the data measured is shown in plot c. Although the number of data points in figure 3.8 c is reduced, the low sensitivity to variations in Re_{jet} for the points exhibiting similar $\frac{\dot{V}_{SS}}{V_{MS}}$ ratios suggests that the side-to-main volume ratio has the largest effect on $\frac{\tau}{t_{mix}}$ and consequently, on the homogeneity of the mixture.

According to the operation parameters for the combustion chamber based on the previous description and summarized in table A.1, the average side-to-main flow ratio during operation is $\frac{\dot{V}_{SS}}{V_{MS}} \approx 0.3$. Hence, it cannot be guaranteed that the constraint $\frac{\tau}{t_{mix}} \geq 10$ is hold during operation. Since a further reduction of the nozzle diameter would result in increasing Re_{jet} but not necessarily improve homogeneity, rising the volume flow being injected, e.g. feeding part of the inert gas together with the air flow, seems to be the most effective alternative to improve the residence-to-mixing time ratio. The need to adopt such measures will

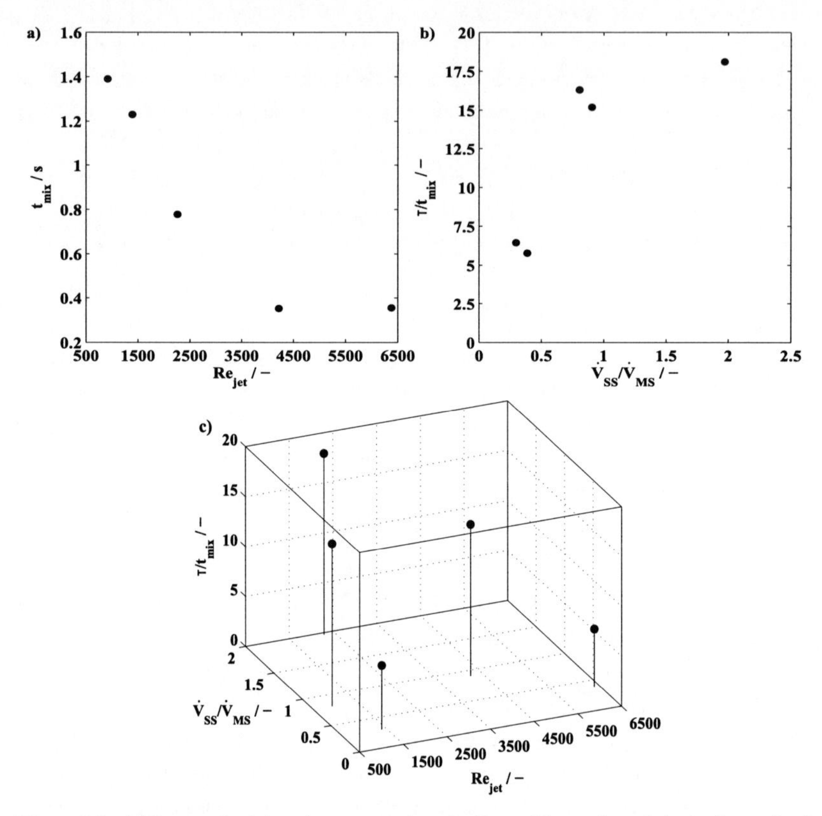

Figure 3.8: a) Measured mixing time t_{mix} against the Reynolds number of the jet for each of the three 1.0 mm injector nozzles Re_{jet} (Setting angle 47°).
b) Residence-to-mixing time relation $\frac{\tau}{t_{mix}}$ as a function of the side-to-main volume flow ratio.
c) $\frac{\tau}{t_{mix}}$ against Re_{jet} and $\frac{\dot{V}_{SS}}{\dot{V}_{MS}}$.

be analyzed based on the analysis of the chamber behaviour under combustion conditions. An injector with three 1.0 mm nozzles has been chosen as most appropriate configuration for this purpose. Since the real time captures of the tracer experiments show a better homogeneity for large setting angles, the direction of the injected flow is chosen to exhibit an inclination of 64.7° (see figure B.1). This configuration is analyzed in detail in combustion experiments described in section 3.4.

3.4 Combustion experiments in a single chamber reactor

After establishment of an appropriate chamber design, its functionality under combustion conditions has been experimentally tested in a single chamber reactor (see figure 3.4). Since the experimental setup strongly resembles the one described in section 5.1 for the RFR under periodic operation, a detailed description at this stage is omitted. It is worth mentioning, that the current setup represents a simplified version of the latter, especially regarding the automation degree.

The current setup is conceived to supply a single chamber reactor with combustion reactants at several loads, i.e. being able to control the composition and flows of fuel, air and inert carrier to simulate different operation conditions. The reactor itself consists of a packed bed, in which a combustion chamber with the geometry described in the previous section is embedded. The tubular reactor is enclosed in a radiation oven that enables preheating the whole system to the desired temperature conditions as well as balancing heat losses during operation.

The experimental work focuses on several aspects of the combustion chamber operation and can be summarized in three sections directly related the issues postulated in section 3.2. The mixing homogeneity studied in terms of tracer methods and reported in 3.3.2 must be validated under combustion conditions. Furthermore, the dynamic of the chamber and its robustness against parameter variations need to be analyzed, as it is intended to be run within a RFR under periodic operation. Last but not least, the fulfillment of the chamber's task, namely the efficient reheating of the neighboring packing at a high temperature level, needs to be experimentally validated. Relevant findings in this respect are reported in the following sections.

3.4.1 Stationary operation

Experimental setup

Proper chamber operation is closely related to the degree of mixing of the gases in it and consequently, to the temperature homogeneity obtained. This is analyzed based on both temperature measurements and visual assessment of the combustion process. For this purpose, the reactor consists of a vertically mounted quartz glass tube with an inner diameter of 50 mm. The bottom of the reactor tube is sealed with a head that enables the introduction of a fuel mixture diluted in an inert gas carrier, mainly steam or nitrogen (see figure B.2). The lower section of the tube is filled with an inert ceramic packing in form of raschig rings, that contributes to uniformly distribute the flow and preheat it up to the desired inlet temperature. On the top of the inert packing, a short section with a height of 2.5 cm is filled with catalytic pellets of 2 mm diameter in average. The chamber itself, consists of an empty section of 50 mm diameter and 50 mm height, delimited by two perforated metallic plates, which are fixed at a central tube that is used to feed the combustion air. The latter is introduced through the injection nozzle described in section 3.3.2, which is positioned in the center of the upper plate. On the top of the chamber, catalytic packing is filled over a length of approximately 10 cm. The top of the quartz glass tube is left opened, so that the combustion off-gases are directly discharged to the atmosphere via an extractor hood and an air blower. The detail of the chamber, both schematically and in the quartz glass reactor can be seen in figure 3.4.

The temperature distribution in the chamber itself and the surrounding packing is monitored with help of a total of 19 K-type thermocouples (0.5 mm) positioned according to figure 3.9.

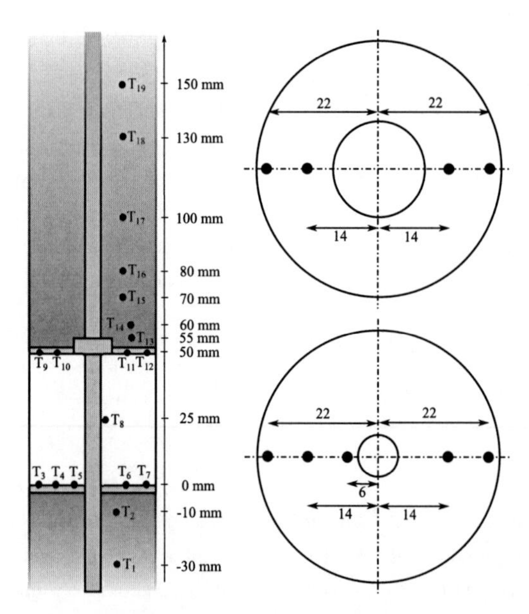

Figure 3.9: Position of the thermocouples in the single combustion chamber quartz glass reactor (left). Right, top: Position within the upper metal plate; right, bottom: position within the bottom plate.

The large amount of sensors used responds to the need to ensure a good resolution of the temperature profiles in the system identified in previous work, in order to detect possible undesired excess temperatures [62, 71]. Thermocouples *T1* and *T2* monitor the temperature of the inlet gas flow and would detect if an undesired combustion below the chamber took place due to undesired penetration of air in the lower packing section. Thermocouples above the chamber are located close to the reactor axis and allow an assessment of the temperature in the packing downstream. Thereby, thermocouples *T13* and *T14* are positioned within a very short distance at the outlet of the chamber, in order to monitor possible temperature maxima, occurring during catalytic combustion of unreacted gases leaving the chamber. Thermocouples *T3* to *T12* measure the temperature distribution in the chamber. While *T8* measures the gas temperature in the geometric center of the chamber, the rest of the sensors measure the temperature distribution along the diameter of the lower and upper perforated metal plates respectively (cf. figure 3.9, right). During the assembly of the latter, special attention is paid to the fact that the alignment of the thermocouple positions in the upper plate mirrors the position of the thermocouples in the lower plate (see figures 3.4 and 3.9). This alignment allows a qualitative assessment of the temperature homogeneity in the chamber by comparison of the temperature profiles along both plates. Furthermore, the choice of the thermocouple positions in the lowest metal plate aims at the detection of the highest and lowest temperatures expected during operation. The highest temperatures are expected in the three regions of the lower plate where the air jets are directed to. Consequently, the lowest temperatures will be measured at each of the positions between the regions where the air jets impact with the lower plate.

Operation conditions

The discussion in the current section analyzes the effect of the most relevant parameters identified with help of tracer experiments, i.e. Re_{jet} and $\frac{\dot{V}_{SS}}{\dot{V}_{MS}}$, on the behaviour of the chamber under combustion conditions. The fuel used is PSA off-gas with the composition given in section 2.1.1. The operation settings are based on the procedure already described in section 3.3.1 and summarized in table A.1. Nitrogen or steam have been used as inert carrier to balance the heat fluxes during production and regeneration. The different operation regimes reported are established by variation of the volume flows and of λ. It should be noticed that, for a given main flow containing both fuel and carrier gas, an arbitrary setting of λ is only feasible within a range defined by $\lambda_{min} \leq \lambda \leq \lambda_{max}$. Below the minimal threshold λ_{min}, energy released by combustion is not enough to reach temperatures at which stable FLOX®

operation can be achieved. Above λ_{max}, temperatures exceeding the temperature limit of $\approx 1000\,°C$ might be reached.

Experimental results

Experimental results under stationary operation are in agreement with the observations made based on tracer experiments and corroborate the suitability of the chamber to attain flameless combustion regimes. However, operation is very sensitive to parameter variations and flame formation cannot be fully suppressed when operating at unfavourable conditions, e.g. low $\frac{\dot{V}_{SS}}{\dot{V}_{MS}}$. Provided that this parameter directly correlates with λ and the latter must be adjusted in order to control the chamber temperature, adverse operation conditions cannot be obviated. Hence, it is of prime importance to detect if, in case of flame appearance, risk of material damage is to be expected.

	Figure a	Figure b
T_{oven} / °C	800	300
$\dot{V}_{MS}$ / $\frac{Nl}{min}$	50.5	50.5
$\dot{V}_{SS}$ / $\frac{Nl}{min}$	14.73	23.75
$y_{N_2,MS}$ / $Vol.\%$	60	60
λ / -	0.31	0.5
$\frac{\dot{V}_{SS}}{\dot{V}_{MS}}$ / -	0.29	0.47
Re_{jet} / -	≈4600	≈7400

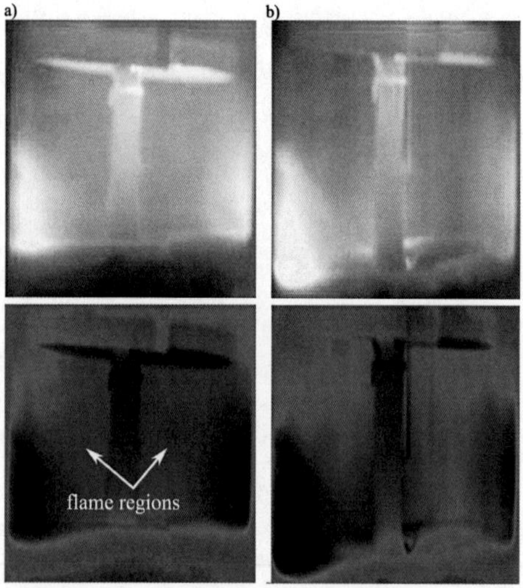

Figure 3.10: Combustion of PSA off-gas diluted in nitrogen with (a) and without (b) perceptible flame regions as a function of the side-to-main volume flow ratio. Top: original pictures; bottom: black light filter applied to the original pictures to enhance the recognition of flame regions.

Since fuel is pre-diluted in the carrier gas, flame formation can only occur in those regions where oxygen is present at high concentrations, namely at the outlet of the injection nozzles. Flames occur to have the form of the air jet and their length depends solely on the mixing degree in the chamber. As soon as the air flow is mixed with the remaining gases in the chamber, flames are not further visible. This corroborates the strong dependency of operation performance on the side-to-main volume flow ratio $\frac{\dot{V}_{SS}}{\dot{V}_{MS}}$ already discussed in section 3.3.2.

The effect of increasing the latter is shown in figure 3.10, where the same amount of fuel/inert is fed to the chamber but combusted to two different extents. In the left picture, λ_{max} has a value of 0.31 since the outer temperature is kept at 800 °C. In the right picture, with an outer temperature of only 300 °C, energy losses are barely balanced and $\lambda_{max} = 0.5$ is required in order to reach analogous average temperatures in the chamber as in the former case. The immediate consequence of increasing λ and thus, the $\frac{\dot{V}_{SS}}{\dot{V}_{MS}}$ ratio, is supression of any perceivable flame region in the left figure.

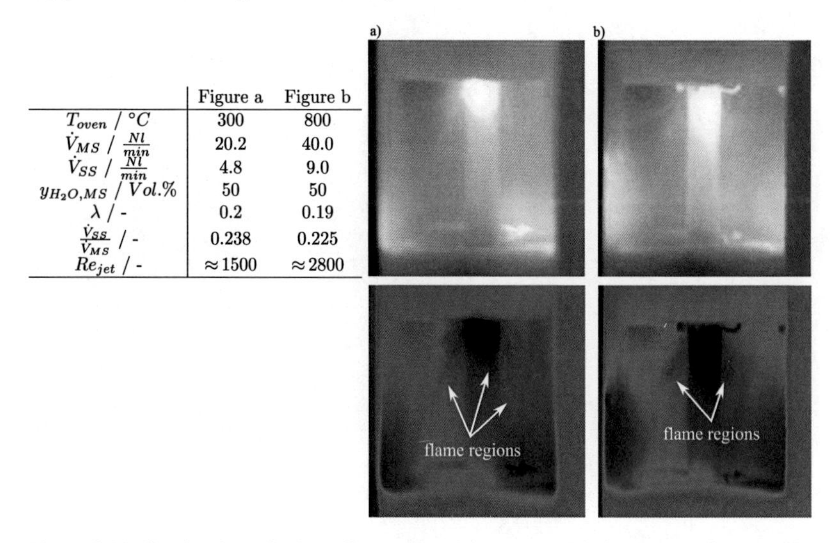

	Figure a	Figure b
T_{oven} / °C	300	800
$\dot{V}_{MS}$ / $\frac{Nl}{min}$	20.2	40.0
$\dot{V}_{SS}$ / $\frac{Nl}{min}$	4.8	9.0
$y_{H_2O,MS}$ / $Vol.\%$	50	50
λ / -	0.2	0.19
$\frac{\dot{V}_{SS}}{\dot{V}_{MS}}$ / -	0.238	0.225
Re_{jet} / -	≈1500	≈2800

Figure 3.11: Combustion of PSA off-gas diluted in steam with formation of perceptible flame regions. Top: original pictures of the chamber under (a) intermediate operation and (b) high operation loads respectively [76]. Bottom: black light filter applied to the original pictures to enhance the recognition of flame regions.

As previously discussed, the momentum of the air jet expressed in terms of Re_{jet} does not seem to be the determining parameter when considering flame suppression. This consideration is reinforced based on the experimental results illustrated in figure 3.11. It shows a chamber under stationary operation with an almost constant $\frac{\dot{V}_{SS}}{\dot{V}_{MS}}$ ratio of ≈ 0.2, which is in the lower range of figure 3.8. The operation load of the chamber in figure b is significantly larger than that in figure a, correlating with an increased momentum of the air jet. Even though increasing the jet velocity induces a better mixing and the percievable dimensions of the flame are reduced, the formation of the latter cannot be avoided. Provided that running the chamber at low $\frac{\dot{V}_{SS}}{\dot{V}_{MS}}$ ratios might be necessary during operation of the RFR, it is important to point out at this stage, that flame formation occurs only in the void section of the chamber and does not endanger the stability of the construction materials.

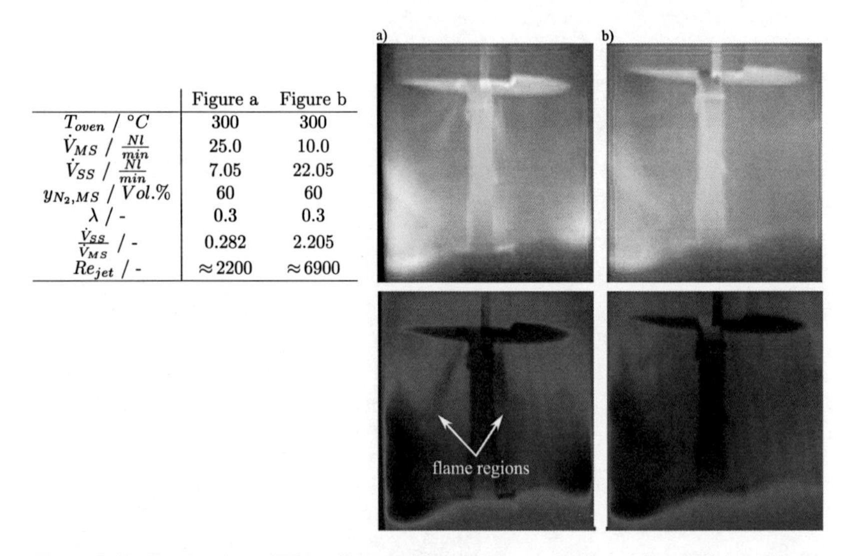

	Figure a	Figure b
T_{oven} / °C	300	300
$\dot{V}_{MS}$ / $\frac{Nl}{min}$	25.0	10.0
$\dot{V}_{SS}$ / $\frac{Nl}{min}$	7.05	22.05
$y_{N_2,MS}$ / $Vol.\%$	60	60
λ / -	0.3	0.3
$\frac{\dot{V}_{SS}}{\dot{V}_{MS}}$ / -	0.282	2.205
Re_{jet} / -	$\approx$2200	$\approx$6900

Figure 3.12: Combustion of PSA off-gas diluted in nitrogen. Left: carrier gas fed as dilutant in the main flow; right: carrier gas fed with combustion air through injection nozzles. Upper and bottom: original pictures and after application of a black light filter to enhance the recognition of flame regions.

Figure 3.11 reveals further interesting information. At low side-to-main volume flow ratios, increasing the total flow rates causes a shift of the hottest zone towards half of the chamber height. This can be seen in the glowing colour of the chamber outer wall in the picture on the right. This effect should be minimized in order to avoid damaging the walls if such operation conditions are prolonged. The problem is obviously minimized if the ratio $\frac{\dot{V}_{SS}}{\dot{V}_{MS}}$ is increased. As already pointed out earlier, the side-to-main flow ratio can be increased if part of the carrier gas is directly added to the air flow. The effect of such an action is shown in figure 3.12, where the sum of all flows and the operation load remains constant, whereas the side-to-main volume flow ratio in figure b is almost one order of magnitude larger than in figure a. Once again, comparing both figures, it becomes obvious that the most effective strategy to suppress flame formation is to increase $\frac{\dot{V}_{SS}}{\dot{V}_{MS}}$.

The conclusions extracted based on the visual assessment of the chamber behavior are also reflected in the temperature distribution within the chamber. This can be seen in figure 3.13, where the radial temperature profiles along the upper and lower plates are shown for the operation states corresponding to figures 3.10 and 3.12.

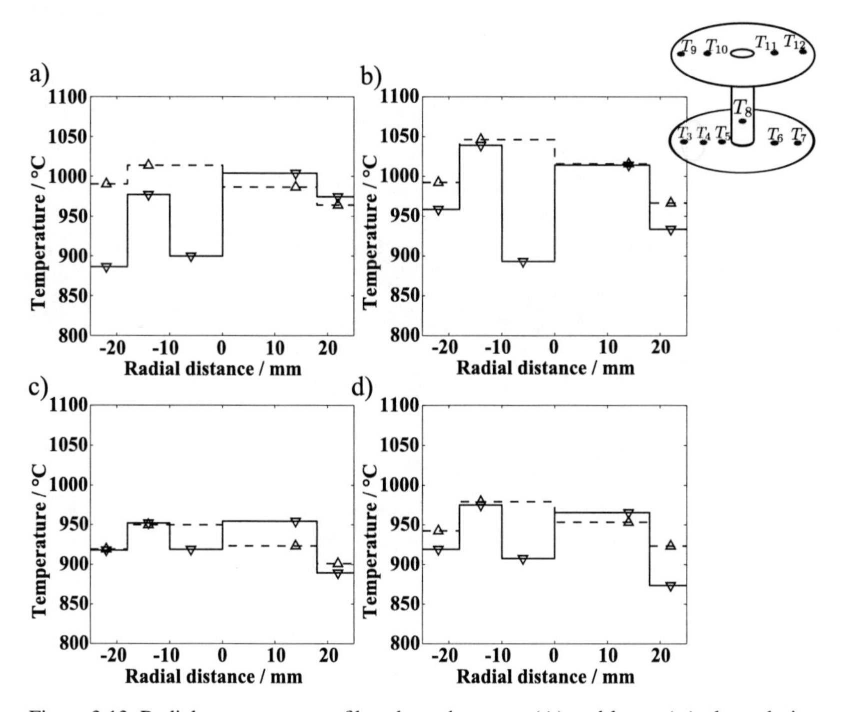

Figure 3.13: Radial temperature profiles along the upper (△) and lower (▽) plates during combustion of PSA off-gas. Plots a and b correspond to the left and right captures in figure 3.10 respectively. Plots c and d to the ones in figure 3.12.

It can be generally stated, that the outermost temperature measurements in both plates show the lowest values and the largest differences, presumably due to the effect of heat losses in the system. At the same time, the information of the thermocouple positioned close to the center of the lower plate (T_5) is not relevant for the analysis of the operation behavior, since the recirculation of the gases at that position is partially hindered by the proximity of the central tube and the temperature measurement is strongly influenced by the cold gases entering the chamber. However, comparison of the temperature measurements T_4 and T_{10} or T_6 and T_{11} (cf. figure 3.9), are a good indicator of the temperature homogeneity in the chamber.

According to figure 3.13, temperature differences between the two plates in the central region of the chamber are larger in those cases where flame formation cannot be avoided (plots a and c), than in those cases where homogeneity is improved and flames are minimized or totally suppressed (plots b and d). Flame appearance can be also interpreted in the somewhat higher temperature of the lower plate compared to that in the upper plate (comparison between plots a and b). In any case, it is remarkable and has to be noticed that even under operation with flame formation (plots a, b and c), temperature differences between both plates are seldom larger than 100 °C. These temperature differences do not represent a relevant drawback for the operation and the mechanical stability of the chamber, provided that the construction is made of high temperature steel. Visual inspection of the chamber after operation does not reveal any remarkable deformation of the chamber internals and corroborates the previous statement.

3.4.2 Ignition and dynamic behaviour

As already introduced in section 3.2, an important aspect to be adressed concerns the variability of the operation parameters of the combustion chamber under reverse-flow operation conditions. Since flameless combustion enables the establishment of a homogeneous, mild temperature that preserves the catalyst and the construction materials from thermal damage, flameless combustion is the preferred operation regime. This, however, implies that the chamber content ignites and reacts homogeneously in the void of the combustion section. Based on the results reported in sections 3.3.2 and 3.4.1, the time required to obtain a homogeneous mixture in the chamber t_{mix} is in the range of seconds. Thus, this should not have a significant effect on the ignition delay of the gas mixture. However, ignition of the flammable mixture just after beginning of the regeneration period is subject to a complex interaction between heterogeneous and homogeneous reactions as well as heat transfer between the different elements and phases in the system. It is therefore important to consider, whether operation of the chamber in an *extinguished* regime represents an alternative during reverse-flow operation.

In the extinguished state, the gas mixture is expected to leave the chamber and react catalytically in the packing downstream whenever the temperature of the flow exhibits values above the catalytic ignition and below the homogeneous ignition point of the mixture. The energy released at the outlet zone of the chamber contributes to heating up the chamber and its content until the homogeneous reaction ignites. This transition, as well as some considerations regarding the possibility to improve the homogeneous ignitability of the mixture are discussed in this section. Prior to this, a brief review of some of the effects playing a role in the interaction between catalytic and homogeneous combustion reactions reported in the literature are briefly summarized.

Previous considerations to the ignition behaviour of the chamber content

The first question that needs to be answered is whether the intended fuel mixture (PSA off-gas) can be catalytically oxidized in case that homogeneous ignition in the chamber does not occur. Therefore, the results reported by Veser et al. in [83] are taken as reference. Veser determined experimentally the heterogeneous and homogeneous ignition temperatures of several hydrocarbons stagnating on a platin surface at several fuel-to-air ratios.

The experimental results are reported as a function of a corrected equivalence ratio Θ, which is defined as

$$\Theta = \frac{\phi}{\phi+1} = \frac{\phi}{\phi+1}$$

ϕ is the so called equivalence ratio and represents the inverse of the air ratio λ previously defined (cf. section 3.2). The advantage of using Θ as parameter to define the fuel/air ratio is that both fuel-rich and fuel-lean mixtures are contained in the range between 0 and 1. $\Theta =$ 0.5 corresponds to a stoichiometric mixture. Thus, $0 < \Theta < 0.5$ refers to fuel-lean mixtures and $0.5 < \Theta < 1$ to fuel-rich ones respectively. Relevant for the current regeneration concept are those fuel-to-air ratios corresponding to $\Theta \geq 0.5$. According to figure 3.14, not only exhibits methane the highest homogeneous ignition from all hydrocarbons, but also the highest catalytic one. However, during cyclic operation of the RFR, temperatures in the packed bed are high enough to ensure the catalytic ignition of the fuel mixture even if it only consists of methane (cf. fig. 2.9).

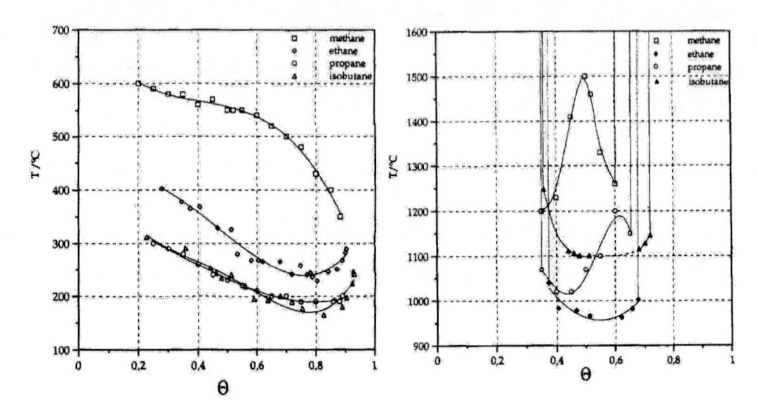

Figure 3.14: Heterogeneous (left) and homogeneous (right) ignition temperatures of several hydrocarbons on a catalytic surface (Pt) as a function of the corrected equivalence ratio [83].

Taking into account that the fuel mixture used during reverse-flow operation contains considerable amounts of lighter ignitable components such as hydrogen, the heterogeneous combustion of the mixture should not present a limitation for the experimental conditions considered. This statement is supported by several studies that sustain the general agreement that the presence of fuels with lower ignition temperatures than methane contribute to widening the flammability limits, lowering the ignition temperature and stabilizing the combustion process. In terms of homogeneous combustion, Jackson et al. report this behaviour in lean premixed CH_4-flames [84], whereas Dagaut et al. do it for combustion in gas turbines under fuel-lean conditions to minimize NO_x emissions [85]. Also in burners operated under flameless oxidation conditions, Derudi et al. have reported the positive effects of hydrogen to help sustain flameless conditions and enhance full hydrocarbon depletion even at lower furnace temperatures and large fuel dilution ratios [69].

Other authors have reached similar conclusions for systems combining heterogeneous and homogeneous combustion. Williams et al., for instance, describe in [86] that the combustion behaviour of mixed fuels exhibits characteristics corresponding to each of the individual fuels contained in it. Their respective influence on the behaviour depends both on the composition of the fuel and on the equivalence ratios at which operation takes place. Also, Scarpa et al. recently reported the contribution of hydrogen to promote methane oxidation in the gas phase, since ignition of the former over catalytic plates activates radical formation reactions that are involved in the homogeneous combustion of the latter [87].

In spite thereof, the picture of the ignition behaviour is strongly affected by the interaction of catalytic and gas phase oxidation reactions. According to the description made by Pfefferle et al. in [88], in the presence of a catalytic surface, combustion occurs mass transfer controlled on the latter, whenever the temperature is above the catalytic ignition point and kinetic limitations can be neglected. This implies that the reaction extent is solely limited by the availability of reactive species reaching the catalytic surface from the bulk gas phase. Under such an operation regime, the temperature of the surface might reach values far beyond the temperature of the gas bulk, enabling that homogeneous reactions become important on the catalytic boundary layer. Whether this effect contributes to the propagation of the homogeneous reaction or the inhibiting effect caused by the presence of catalytic (at lower temperatures) and even non-catalytic (at higher temperatures) surfaces prevails, remains unclear.

This effect finds several explanations in the literature. According to Vlachos et al. [89], sensitivity of the ignition and extinction behaviour of a flammable mixture decreases with increasing temperature. If temperature is not high enough, formation of radicals through homogeneous reactions might not be fast enough in order to ensure that large concentrations of such species are available for further reaction. Vlachos also points out that under fuel-rich conditions, as in the current case, termination reactions might play an important role contributing to the extinction behaviour of the mixture [90]. Similarly, the presence of catalytic surfaces might further contribute to a depletion of radicals, inhibiting the propagation of homogeneous reactions as well as increasing the homogeneous ignition temperature due to the adsorption of O_2 on the catalytic surface or due to the catalytic formation of water [83, 90, 91]. As already mentioned, non-catalytic surfaces might have a similar effect if surface temperatures are high enough [86, 91].

Given the obvious difficulty to directly transfer the results reported in the literature to the system of study, an analysis of the ignition behaviour in the chamber based on experimental results is necessary. The results are briefly described in the following.

Transition from heterogeneous to homogeneous combustion

The great advantage of the behaviour described above is the robustness of operation. Even if the temperature of the bulk phase is well below the homogeneous ignition temperature, the temperature in the upper part of the chamber increases considerably through heterogeneous combustion of the gas mixture. This behaviour is depicted in figure 3.15, where each curve represents the temperature evolution of several thermocouples located at different positions whithin the chamber and the surrounding packing after supply of a flammable mixture (cf. fig. 3.9).

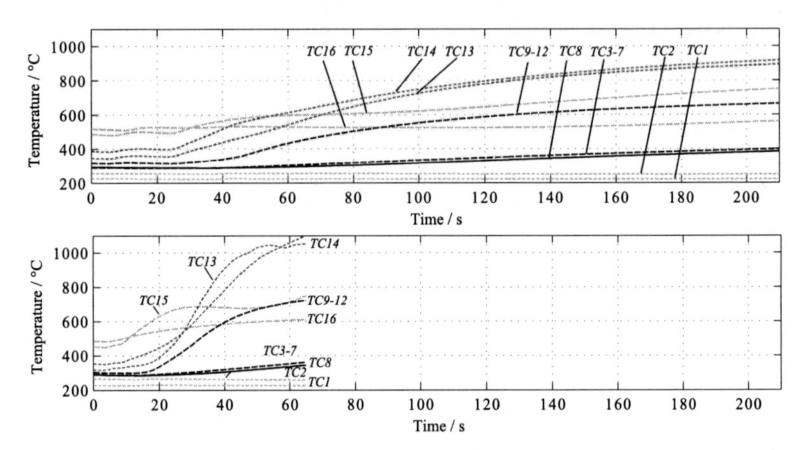

Figure 3.15: Temperature evolution after supply of a flammable mixture in an extinguished combustion chamber. Combustion under stoichiometric conditions ($\lambda = 1$) for 2 Nl/min (top) and 5 Nl/min (bottom) PSA off-gas. The position of the thermocouples TC_i is given in figure 3.9. Adapted from [76].

Whereas the temperature in the inlet region of the chamber (*TC1* and *TC2*) remains invariable during the whole experiment, the temperature in the catalytic packing at the outlet exhibits the highest values measured (*TC13* and *TC14*). The averaged temperature of the upper plate (*TC9-12*) increases proportionally due to heat conductivity while the temperatures in the chamber itself (*TC8* and the averaged temperature in the lower plate *TC3-7*) exhibit a slower increase due to radiation and convective energy transport caused by the recirculating gases. The temperature in the packing further downstream (*TC15* and *TC16*) shows an interesting behaviour. At the beginning of the experiment, the latter is slightly higher than the temperature at the chamber exit (*TC13* and *TC14*). Since the combustion kinetics are enhanced at higher temperatures, the main reaction zone stabilizes further downstream in the catalytic packing. Heat release and conduction, but presumably also partial reaction in the somewhat colder zone origins an upwind displacement of the reaction zone towards the chamber outlet. Thus, after a short period, the temperatures measured by *TC13-14* exceeds the ones measured by *TC15-16*. The latter decreases slightly as the reaction zone migrates upstream, followed by an increase as the thermal wave generated in the position of *TC13-14* is transported convectively in the flow direction.

Using PSA off-gas as fuel, it can be expected that the presence of hydrogen contributes to reducing the homogeneous ignition temperature of the mixture. As soon as the fuel reacts catalytically, it is possible to reach temperatures significantly above the homogeneous ignition threshold in the upper part or the chamber (see *TC9-12* in fig. 3.15). However, the experimental results suggest that exceeding the ignition temperature in the outlet region of the chamber does not suffice to promote homogeneous reaction in the bulk of the gas phase or at least, not to a significant extent. This observation is summarized in figure 3.16. The latter represents temperature information during stationary operation in extinguished (figure a) and ignited (figure b) modes. The flow regime in both figures is identical, so that the only difference between them is the preheating temperature of the system up to 300 and 800 °C respectively.

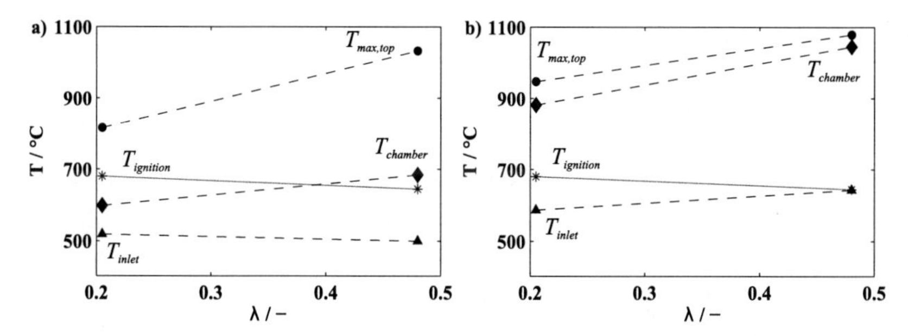

Figure 3.16: Temperature distribution in the chamber for heterogeneous (left) and homogeneous (right) combustion regimes for 20.2 Nl/min PSA diluted in 30.3 Nl/min N_2 (s. table A.1). $T_{max,top}$ is the maximal temperature in the catalytic packing at the chamber outlet; $T_{chamber}$ is the temperature of the gas bulk (*TC8*); T_{inlet} is the temperature at the chamber inlet (*TC2*). $T_{ignition}$ represents the theoretic adiabatic ignition temperature of the mixture.

The temperatures at different chamber positions (top, gas bulk and inlet) are plotted against the theoretical adiabatic ignition temperature of the mixture $T_{ignition}$[4]. Interestingly, even if the temperature in the upper zone of the combustion chamber is well above the homogeneous ignition temperature, combustion in the gas bulk does not occur or it occurs to

[4]The estimation is made based on the combustion reaction mechanism GRI 3.0. The model used corresponds to that of an adiabatic continuously stirred tank reactor (CSTR) operated at the same conditions as the experimental runs. The ignition temperature is assumed to be the highest temperature in the extinguished bulk prior to the transition into an ignited state.

a very low extent if the inlet temperature of the gas flow is significantly below the ignition temperature (cf. figure 3.16, left). Although several aspects, such as heat losses, might affect this behaviour, it can be stated that under reverse-flow operation, only if the inlet temperature of the chamber is close or above the adiabatic ignition temperature of the mixture, homogeneous ignition can be expected.

Based on this observation, the possibility to promote homogeneous ignition of the chamber content by introducing a noble metal catalyst or using of a light flammable mixture are next reported.

Effect of a catalyst to promote homogeneous combustion

The introduction of a platinum gauze, rolled around the central tube of the chamber, represents an attempt to enhance the ignitability of the gas mixture. However, the negligible consequence of this action can be seen in the results plotted in figure 3.17. The latter represents, for two different preheating temperatures, the axial temperature profiles along the chamber and the packing below and above it. The only perceivable difference between the temperature profiles with and without catalytic gauze in the chamber is the temperature measured in the geometric center of the chamber, at position 25 mm (*T8* in fig. 3.9).

At a preheating temperature of 500 °C, the temperature of the inlet flow is well below the homogeneous ignition temperature of the mixture. Thus, operation resembles the behaviour already depicted in figure 3.16, left, where combustion occurs mainly catalytically in the packing at the chamber outlet. The increased reaction extent in the gas phase promoted by the presence of the catalytic surface in the chamber, observable in the almost 100 °C difference in the middle of it, does not significantly contribute to force the transition from a heterogeneous to a homogeneous combustion regime. Thus, the position of the main reaction zone remains at the outlet of the combustion chamber in flow direction. A further interesting observation is the fact that the temperature profiles with and without catalytic support are almost equal except for the temperature in the middle of the chamber. Since the position of the thermocouple exhibiting the largest temperature difference is very close to the platinum mesh, this results suggest that the deviation of both profiles is rather caused by the local measurement of a region where surface reaction takes place, than by a noteworthy reaction in the gas bulk promoted by the catalytic surface.

At higher preheating temperatures (figure 3.17, right) homogeneous reaction occurs predominantly, since the inlet temperature approaches the adiabatic ignition temperature of the mixture and the heat losses of the system are balanced with help of the radiation oven. As in

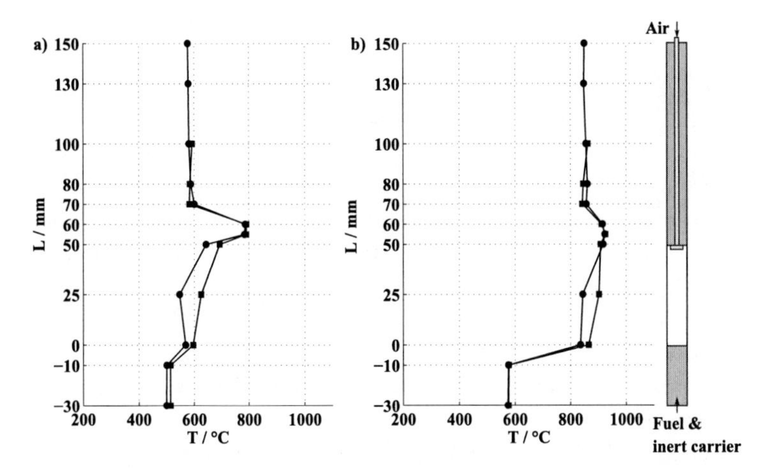

Figure 3.17: Temperature distribution in the chamber and the surrounding packing for combustion with (■) and without (•) a Pt-gauze in the chamber. Radiation oven at two preheating temperatures: a) 500 °C and b) 800 °C. Combustion of 20.2 Nl/min PSA diluted in 30.3 Nl/min N_2 and $\lambda = 0.206$. Temperature measurement according to figure 3.9.

the previous case, the similarity of both temperature profiles suggests that the difference of approximately 50 °C is not representative for the temperature of the gas bulk and correlates with the proximity of the thermocouple and the platinum gauze around the chamber axis.

The small contribution of the catalytic surface on the homogeneous ignition observed can be considered to be in good agreement with the observations reported in the literature and discussed previously. The platinum gauze does not seem to have an inhibiting effect on the propagation of the homogeneous reactions, but it does not boost it in a significant manner either. The depletion of radicals at the temperature level of the chamber could be a reason therefore, although the positioning of the gauze could also play a role. Taking into account, that the reaction on the catalyst surface is mass transfer limited, the catalyst should be better positioned in a region where unreacted fuel and oxygen are predominantly found. Thus, the performance of the catalyst could be potentially improved if it could be positioned at the outermost diameter of the chamber. This measure, however, has been not adopted for the following studies to characterize the performance of the combustion chamber.

Ignition with light flammable mixtures

Based on the experimental findings discussed above, a robust, spontaneous ignition of the homogeneous combustion is only to be expected if the temperature of the inlet flow is close enough to the adiabatic ignition temperature of the fuel mixture. Since the inlet flow temperature during periodic operation of the RFR cannot be directly influenced and the presence of a catalytic surface in the chamber does not significantly contribute to accelerate transition between heterogeneous and homogeneous combustion, a further alternative has been considered. Eichhorn systematically tested under which operation conditions (i.e. fuel mixture and flows) homogeneous ignition occurred within a short time interval after the begin of the regeneration step. The results reported in [76] suggest that the use of a fuel with low ignition temperatures (e.g. hydrogen) is required for a fast ignition if high inlet temperatures are not to be expected. The energy release in the chamber, proportional to the volume flow, must be high enough to counteract the heat losses of the system. At the same time, the maximal flow needs to be limited in order to guarantee sufficient residence times in the chamber and limit the cooling effect of the gas entering the chamber. Figure 3.18 summarizes some of the results, which show the ignition delay of several H_2-N_2-mixtures under stoichiometric conditions, depending on the H_2 and N_2 flows. This delay is defined as the time interval between the temperature rise on the upper region of the chamber due to heterogeneous combustion and the temperature increase recorded in the bulk as soon as the homogeneous reaction ignites. The initial temperature of the system is 200 °C.

If the operation conditions are properly choosen, relatively short ignition delays (below 25 s) can be achieved. Considering that the initial temperatures during periodic operation of the RFR are significantly higher than 200 °C, spontaneous ignition of the chamber content could be expected. One possibility to operate the reactor based upon this observation consists of igniting the chamber with a fuel mixture designed for this purpose and dynamically switching to a conventional fuel, as soon as the temperature in the system ensures that homogeneous combustion of the mixture can be sustained. However, the obvious apparative and operative complexity of this approach highlights the need to develope and run the regeneration period based on a simple strategy.

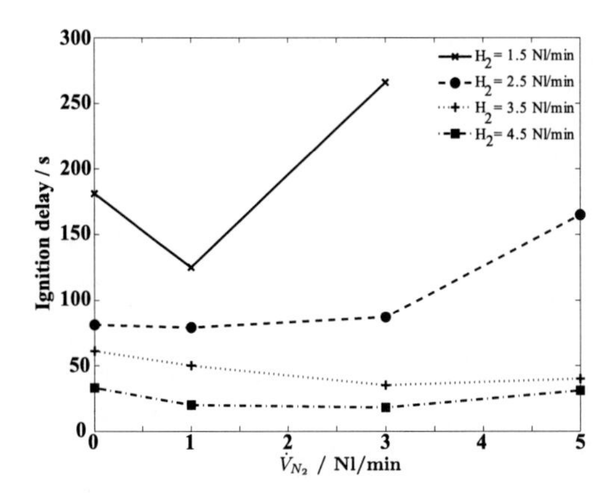

Figure 3.18: Ignition delay of a H_2-N_2-mixture as a function of the H_2 and N_2 volume flows ($\lambda = 1$). The chamber is initially preheated at 200 °C [76].

Remarks on the ignition behaviour of the chamber

The results discussed in the current section confirm the possibility that the chamber does not spontaneously ignite at the beginning of the regeneration period. Although homogeneous combustion is the preferred operation regime, the attempt to enhance the ignition behaviour with simple strategies, such as the introduction of a catalytic surface in the chamber, does not show a significant improvement. In spite thereof, the experimental results confirm the potential to operate the chamber in *extinguished* state. Provided that the mixing performance of the chamber remains unchanged, a homogeneously mixed reactive flow leaves the chamber and combustion takes place catalytically in the active packing downstream. Homogeneity of the gas flow is a key aspect, since it enables a uniform reaction without appearance of hot-spots. Thus, this operation mode represents a robust alternative whenever the gas bulk does not immediately ignite. The consequences of this form of operation are next discussed.

3.4.3 Understoichiometric combustion - Revision of the regeneration concept

During periodic operation of the RFR, each of the combustion chambers embedded in the packed bed is to be operated under fuel rich conditions. The experimental results reported in the previous sections 3.4.1 and 3.4.2 are based on this premise. However, a closer look to the temperature profiles in the packing downstream of the chamber reveals an undesired effect (see figure 3.17, left). Oxidation of the fuel occurs primarily in the catalytic zone that expands along the first $10 - 15\,\mathrm{mm}$ after the chamber, where the highest temperatures are measured. After depletion of the supplied oxygen, a notorious temperature decrease occurs, so that the temperature established in the catalytic packing downstream is far below the maximal temperature reached in the reaction zone. This behaviour occurs whenever the methane contained in the fuel is not completely reacted in the combustion chamber or the catalytic reaction zone donstream.

The temperature profile resembles those established during autothermal reforming [92] or selective oxidative dehydrogenation of hydrocarbons, where the oxidation reaction occurs at the entrance of the catalytic zone and the endothermic reaction does not occur until the oxidizing component is depleted and the temperature is sufficiently high to overcome the kinetic limitations to which the endothermic reaction is subject. In the current case, the large methane concentration in the synthetic PSA off-gas used (12.6%) suggests that reforming reactions are the main cause for the above mentioned effect. However, the reaction mechanism behind is not unambiguosly defined. Several studies focused on the catalytic combustion of methane, its oxidative coupling or even the production of synthesis gas, reveal information about possible reaction pathways and their dependance on the operation conditions, which can be applied to the current case.

Hickman et al. [93] studied the influence of the metal coating of a monolith used for the production of synthesis gas by direct oxidation of methane. They reported higher hydrogen and carbon monoxide selectivities for Rh-coated monoliths than for Pt-coated ones, as well as improved selectivities in case of using oxygen instead of air as oxidizing agent. Furthermore, they observed that both the autothermal temperature rise and the residence times might play an important role in the reaction pathway to form H_2 and CO. At low autothermal temperatures and short residence times, the latter are mainly formed through direct oxidation reactions, since reforming reactions require the previous formation of H_2O and CO_2 and thus, higher autothermal temperatures or larger residence times [93]. Tsang et al. [94] also describe the competition between the two mechanisms, *CRR* and *DPO*, in the formation of synthesis gas from methane. The first stays for *Combustion and Reforming Reactions mechanism*, whereas the second stays for *Direct Patial Oxidation mechanism*. Their results, in agreement with those from Hickman et al., reveal that both mechanisms might describe correct reaction paths depending on the operation conditions. Thus, for short residence times, *DPO* might be the prevailing mechanism, since total oxidation and reforming reactions require longer residence times in order to reach thermodinamic equilibrium [94]. At this stage, it is also worth mentioning that further reaction mechanisms could also play a role. Specially if temperatures are high enough, homogeneous reactions (combustion as well as oxidative coupling of methane) might have to be considered.

Another aspect that should be taken into account, is the fact that besides methane, the synthetic PSA off-gas used for the experimental work contains considerable amounts of hydrogen, carbon monoxide and dioxide and large amounts of inert components introduced together with the fuel and with the combustion air. It has been reported that diverse noble metals exhibit different activities for the catalytic oxidation of methane, being Pt/Al_2O_3 and Rh/Al_2O_3 the less active ones [95]. Hence, it could be also considered that some of the flammable components present in the fuel mixture might be selectively oxidized, remaining relatively large amounts of unreacted methane that can undergo endothermic reforming.

The cooling effect of the after-reaction is especially pronounced for those operation states at which homogeneous reaction in the chamber does not occur or only at a low extent (cf. figure 3.17). The observation of Scarpa et al. [87], who studied the effect of combusting methane in presence of hydrogen, suggests that larger homogeneous reaction extents should contribute to minimizing the endothermic after-reaction observed in the catalytic packing downsteam. Their results, in agreement with those from many other authors, point out that hydrogen promotes the oxidation of methane since above 700 °C the production of radicals H, O and OH, activated by the ignition of hydrogen, enhance the homogeneous reaction of methane. The preferential reaction of these radicals with methane has an inhibiting effect on the homogeneous hydrogen reaction [87]. Thus, in case of significant reaction in the chamber (e.g. figure 3.17, right), the amount of unreacted methane undergoing an endothermic reaction in the catalytic packing downstream is reduced and consequently the temperature decrease is damped.

This may not hold true for temperatures below 700 °C in the chamber, since homogeneous hydrogen combustion occurs preferentially. In an attempt to quantify the reactants depletion through homogeneous reaction in the chamber, series of combustion experiments including a chromatographic analysis of the gas composition at the entrance of the combustion chamber, in the chamber, and after 10 cm of catalytic packing were performed. A platin gauze was introduced in the chamber in order to enhance the homogeneous reaction. The results summarized in figure 3.19 show that at low air-to-fuel ratios ($\lambda < 0.5$) significant amounts of methane are present despite the homogeneous reaction in presence of hydrogen. Furthermore, the increase of hydrogen and carbon monoxide content at the oulet of

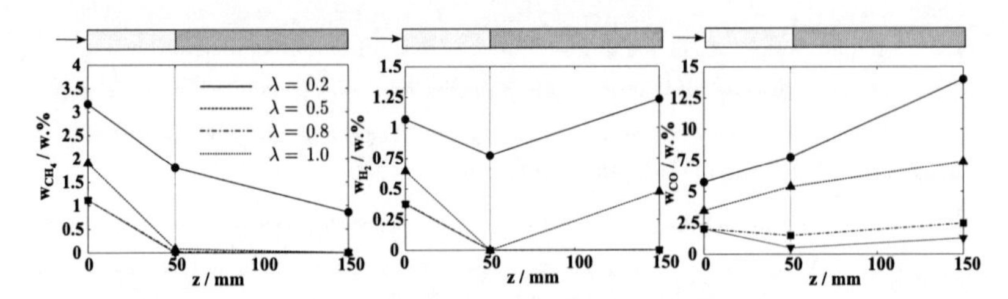

Figure 3.19: Mass fractions of CH_4 (left), H_2 (middle) and CO (right) in the feed, in the chamber and after the catalytic packing during combustion of 20 Nl/min PSA off – gas diluted in 20 Nl/min N_2 at several fuel-to-air ratios.

the catalytic zone for low air-to-fuel ratios suggests that steam reforming of methane might be indeed responsible for the temperature decrease in the catalytic packing. For $\lambda > 0.5$, both hydrogen and methane are almost fully depleted in the combustion chamber, being carbon monoxide the only oxidable component detected at the oulet of the catalytic packing. An exhaustive analysis of the reaction mechanisms involved is far beyond the scope of this work. The results, however, serve as evidence that homogeneous ignition of the gas mixture in the chamber does not suffice to ensure the supression of an endothermic after-reaction under fuel rich conditions and hence, alternatives to avoid the efficiency losses in the energy supply need to be developed.

Implications of understoichiometric combustion

The first conclusion deriving from the combustion results under fuel rich conditions is that efficiency losses must be taken into account, whenever endothermic reactions can occur in the catalytic packing downstream from the combustion chamber. According to the observations reported, several aspects such as the combustion regime (homogeneous or heterogeneous), fuel composition, air-to-fuel ratio λ or even the inert gas carrier used (N_2 or H_2O) play thereby a role. Thus, it becomes obvious that a systematic control of these parameters to minimize the effects of the undesired reactions is an arduous task.

Besides the reheating efficiency, one of the core elements of the regeneration strategy must be a robust control scheme that enables to adapt the heat released in each of the chambers embedded in the packed bed, in accordance to the changing temperature conditions along the reactor. Taking the reheating strategy represented in figure 2.6 as reference, the total amount of fuel required to reheat the packed bed is supplied to one of the reactor ends.

In each of the several combustion chambers connected in series, only a portion of the fuel is reacted ($\lambda_i < 1$). Thus, the total amount of air being supplied to all the combustion chambers should fulfill

$$\sum_{i=1}^{I} \lambda_i = 1,$$

while the value of λ_i must be continuously adjusted according to the temperature and composition changes of the flow. These composition changes, originated by the extent to which combustion occurs in each chamber as well as by the endothermic parallel reactions taking place in the packing downstream, motivate reconsidering the current reheating strategy. Hence, a simplified concept is proposed and analyzed in the following chapter.

3.5 Summary

The proven technology of flameless combustion FLOX® sets the basis for a novel reheating strategy for the RFR introduced in chapter 2. The transfer of the industrial FLOX® burners operation principle to a chamber of reduced dimensions embeded in a packed bed is accompanied by several technological challenges that have been analyzed based on a set of different experiments.

The applicability of the concept and the possibility to generate large gas recirculation ratios in the combustion chamber has been studied with CFD simulations, whereas the definition of the design details and the validity of the concept have been verified with help of tracer methods. The latter served to identify the side-to-main flow ratio $\frac{\dot{V}_{SS}}{\dot{V}_{MS}}$ as key operation parameter, correlating with a rapid and homogeneous mixing of the chamber content. The sensitivity of the chamber behaviour towards changes in $\frac{\dot{V}_{SS}}{\dot{V}_{MS}}$ has been validated based on combustion experiments in a combustion chamber.

Although flameless combustion can be established, the changing conditions during the regeneration step may cause that flame formation cannot be suppressed under certain operation regimes. In spite thereof, the experimental results have shown that even in this case, the flame region is locally delimited and represents no significant risk of material damage for the reactor and the catalytic packing.

A further important observation deriving from the experimental study is the ignition behaviour of the chamber content, which has proven to be strongly dependent on the temperature and composition of the inlet flows. Fortunately, homogeneous combustion is not mandatory, since the experimental results have also proven the potential to operate the chamber in *extinguished* regime. Under this operation mode, the chamber acts as mixing unit and the gas mixture reacts uniformly in the catalytic packing downstream, without appearance of undesired hot-spots. Thus, safe operation is possible even if the gas bulk does not ignite or ignites far after the beginning of the regeneration period.

In contrast, occurrence of undesired endothermic reactions in the catalytic packing have been identified as the cause of efficiency losses, as well as representing a hurdle in the definition of a robust control strategy for the reheating process. These reactions are a direct consequence of the understoichiometric combustion strategy pursued so far. Consequently, the work reported in the following chapter reconsiders the regeneration strategy and proposes an alternative that overcomes several of the drawbacks identified and reported in this chapter.

Chapter 4

Reheating strategy

The results reported in chapter 3 verify the potential of the combustion chamber as a promising concept to reheat a packed bed by *in situ* combustion. However, the reheating strategy pursued so far, based on understoichiometric combustion of a fuel mixture fed to one reactor end, disclosed several operative drawbacks that suggest the need to reconsider the regeneration concept. Focus of the current chapter is the postulation and validation of a reheating strategy based on stoichiometric combustion of a well defined fuel mixture in each of the combustion chambers. The implications in the reheating concept and the operation mode, as well as in the design of the chambers are next reported.

4.1 Stoichiometric combustion concept

According to the results discussed in the previous chapter, stoichiometric combustion in each of the chambers represents an appropriate approach to successfully suppress after-reaction in the catalytic packing, even if ignition of the homogeneous reaction in the chamber does not take place. In such a case, the chamber can be considered as a mixing unit that enables catalytic combustion of a homogeneous mixture in the packing without formation of hot-spots. Furthermore, such a configuration grants an extraordinary robustness of operation, which could not be guaranteed with operation under fuel rich conditions.

Robustness is mainly granted by the autonomy that stoichiometric operation confers to each of the several chambers that might be connected in series along the packed bed. Since fuel and oxidizing agents react completely, an inert flow leaves each of the chambers. This allows for a considerable simplification of the control strategy required under reverse-flow operation.

Operation under stoichiometric conditions requires a design modification in order to enable independent fuel supply to each of the chambers (cf. fig. 4.1). Since the fuel volume supplied represents only around 10% of the total volume flow, injection from the side of the chamber and perpendicular to the main flow direction should not affect the mixing process and has been chosen as design concept. Direct consequence of this decision is that the combustion chamber and the reactor itself must be constructed with high temperature steel, in order to allow the introduction of side-feeds (see figs. 5.2 and 5.3). Although the fuel supply concept offers room for improvement (cf. sec. 4.3), the configuration in figure 4.1 serves as experimental proof-of-concept of the new strategy.

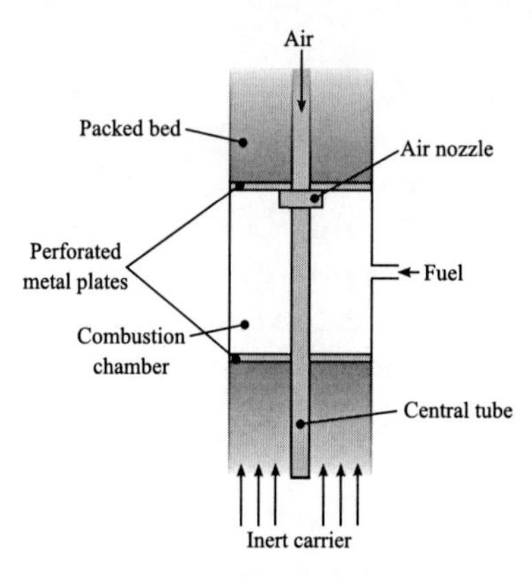

Figure 4.1: Representative configuration of a flameless combustion chamber with independent fuel supply to enable stoichiometric combustion (cf. fig. 3.2).

Comparison of figure 3.17 in the previous chapter and figure 4.2, reveals the essential difference in the reheating efficiencies for understoichiometric and stoichiometric combustion conditions respectively. It is worth noticing that the inlet flow and the chamber bulk temperatures in figure 4.2 are well below the homogeneous ignition temperature of the mixture. Thus, reaction in the chamber occurs only to a marginal extent (cf. fig. 4.2, left) and the operation regime can be considered as *extinguished* according to the previous definition of this term. Despite this fact, endothermic after-reactions can be fully suppressed and the temperature decrease along the packed bed is solely attributed to heat losses. The latter are influenced by the preheating temperature of the radiation oven, explaining the differences between both profiles in figure 4.2.

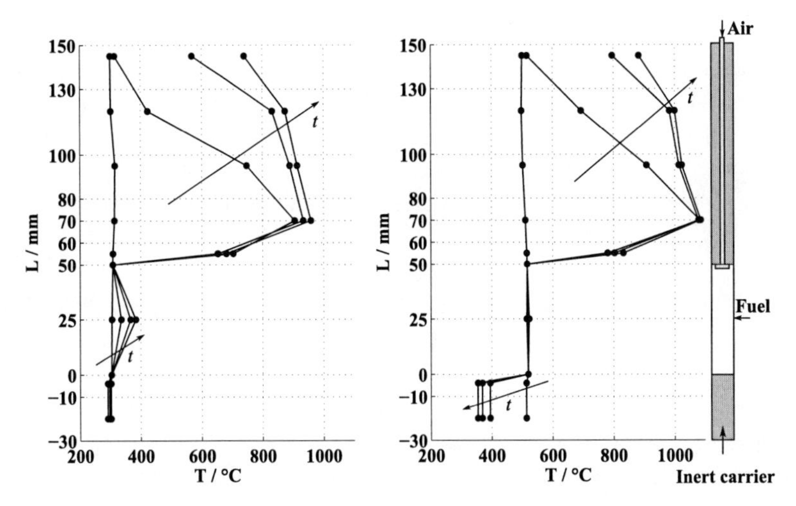

Figure 4.2: Axial temperature evolution for stoichiometric fuel-to-air supply without homogeneous combustion in the chamber. Profiles from $t = 0$s to $t = 300$s every 100s. Radiation oven at two preheating temperatures: a) 300°C and b) 500°C. Combustion of 6.7Nl/min PSA diluted in 36Nl/min N_2 and $\lambda = 1$ (cf. tab. A.1).

4.2 Design constraints and combustion chamber operation range

As already mentioned, a regeneration approach with multiple chambers connected in series implies that each of them must be provided with an individual fuel supply. This requires to adjust the fuel and air flows as a function of the inlet conditions at each of the combustion chambers. This consideration is concretized in the following subsection.

4.2.1 Operation constraints

The need to introduce an inert component during the reheating step in order to balance the heat fluxes between production and regeneration is also required in the current regeneration approach. However, in contrast to the strategy based on understoichiometric combustion, the main flow during the reheating step consists solely of the inert carrier gas, whereas fuel and air are individually supplied to each of the chambers, which can be ideally described as adiabatic stirred reactors. Hence, the temperature difference between inlet and outlet flows is explicitely related to the energy released by combustion and is directly proportional to the fuel and air flows supplied at each moment. Under assumption of stationary, perfectly adiabatic operation conditions, the temperature difference between inlet T^+ and outlet T_{out} can be defined in terms of the adiabatic temperature increase ΔT_{ad}, which reads

$$\Delta T_{ad} = T_{out} - T^+ = \frac{\sum_{i=1}^{I} \dot{\xi}_i (-\Delta h_{C,i})^+}{\sum_{j=1}^{J} \dot{m}_j \cdot \overline{c_{p,j}}\Big|_{T^+}^{T_{ad}}} \tag{4.1}$$

where $\dot{\xi}_i$ represents the reaction extent per unit time and T_{out} is the desired temperature level, which the packed bed is to be reheated to. Under assumption of complete, spontaneous combustion of the fuel entering the chamber, the term in the numerator of the equation can be approximated as

$$\sum_{i=1}^{I} \dot{\xi}_i (-\Delta h_{C,i})^+ \cong \dot{m}_{fuel,theor} \sum_{i=1}^{I} \frac{w_i}{MW_i} (-\Delta h_{C,i})^+ \tag{4.2}$$

with the subscript i referring to the combustible species H_2, CO and CH_4 contained in the fuel. Since the inlet temperature in each chamber is measured during operation, equation 4.1 can be reformulated to estimate the theoretical amount of fuel required at each moment, the latter being a function of following parameters

$$\dot{m}_{fuel,theor.} = f\left(T_{out}, T^+, \dot{m}_j, w_i\right), \tag{4.3}$$

where the subscript j refers to the inert components being heated up in the process. Further dependencies between the flows entering and leaving each combustion chamber, and in particular the fuel-to-air ratio, are given by the stochiometry of the combustion reactions considered (cf. sec. 2.1.1).

For a combustion chamber operating under the above mentioned conditions and using both PSA off-gas (cf. tab. 2.1) and H_2 as fuel, a map of Re_{jet} and the side-to-main volume flow ratio can be generated as a function of the inlet temperature and the inert volume flow, as shown in figure 4.3. The inert volume flow -also referred as operation load- ranges from 10 to $40\,\mathrm{Nl/min}$, in agreement with the operation parameters considered as relevant for

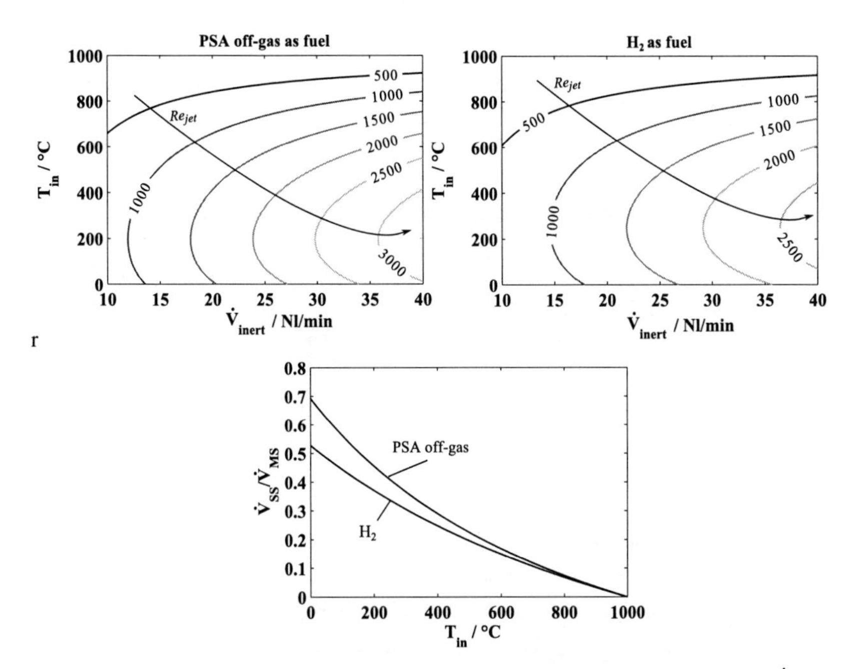

Figure 4.3: Top: Contour lines representing the value of Re_{jet} as a function of T_{in} and $\dot{V}_{inert}$ for stoichiometric combustion of PSA off-gas (left) and H_2 (right). Bottom: Resulting $\frac{\dot{V}_{SS}}{\dot{V}_{MS}}$ ratio against T_{in} assuming stoichiometric combustion conditions.

operation in the RFR (cf. tab. A.1). The plots on the top represent a mapping of Re_{jet} corresponding to the air volume flow required to stoichiometrically oxidize the fuel supplied against the inert volume flow $\dot{V}_{inert}$. As discussed in chapter 3, the side-to-main volume flow ratio $\frac{\dot{V}_{SS}}{\dot{V}_{MS}}$ is a key indicator of the mixing quality and of the temperature distribution in the chamber in case that homogeneous combustion occurs. The bottom plot in figure 4.3 represents the evolution of this ratio as a function of the inlet temperature of the flow, which is the only parameter affecting the former magnitude. Interestingly, already if T_{in} exceeds values of $\approx 400\,^{\circ}$C in the case of PSA off-gas and somewhat lower in that of H_2, the $\frac{\dot{V}_{SS}}{\dot{V}_{MS}}$ ratio falls below a threshold of 0.3. Combustion experiments with $\frac{\dot{V}_{SS}}{\dot{V}_{MS}}$ beneath this value have previously shown that flame suppression cannot be granted (cf. sec. 3.4.1). Although previous results have provided enough evidence that this is not necessarily a synonym of a detrimental operation regime, flames should be obviously avoided. In order to corroborate whether the current reheating approach could represent a drawback during reverse-flow operation, combustion experiments within the $\dot{V}_{inert}$ and T_{in} range depicted in figure 4.3 have been run and the chamber performance characterized. The results are next discussed.

4.2.2 Chamber performance within the design operation range

Aim of the experimental work reported here is to delimitate the operation parameters range, for which the desired homogeneity in the chamber can be achieved. Important indicators thereby are temperature uniformity (cf. sec. 3.4.1) and fuel conversion in the chamber. In the current experiments, additionally to the temperature measurements located in analog positions to those depicted in figure 3.9, the bulk of the gas phase is monitored with two thermocouples located at 20 and 40 mm from the chamber bottom respectively (previously only thermocouple *T8* according to fig. 3.9). The difference between these two measurements under stationary operation, ΔT_{bulk}, serves as indicator of the temperature homogeneity in the gas bulk. Additionally, the maximal temperature gradient within the lowest region of the chamber, ΔT_{plate}, is taken as further reference indicator.

Data reported has been obtained within extensive series of experiments in a single chamber reactor reported by Klump in [96]. Contrary to most of the results previously described, obtained in a single chamber reactor heated up with a radiation oven, the current setup is insulated in order to resemble operation in the RFR (cf. section 5.1). The inert volume flow is varied between 10 to 40 Nl/min and the feed can be preheated up to temperatures between 550 and 650 °C, depending on the flow rate.

Amongst others, the results reveal relevant information regarding the ignition temperature of the homogeneous combustion reaction. Both H_2 and PSA off-gas have been used as fuel for the study. The former ignites at temperatures in the range of 590 – 650 °C, while the latter ignites when temperatures close to 790 °C are measured in the chamber. This has been observed only for inlet temperatures beyond 580 °C and is in full agreement with the results already discussed in figure 3.16, corroborating the observation that the inlet temperature plays a major role in the combustion regime established in the chamber during regeneration.

The current discussion focuses on operation under homogeneous combustion. The establishment of this regime for the whole operation range was only possible with H_2 as fuel. Accordingly, and in spite of the poorer mixing quality expected in comparison to operation with PSA off-gas (cf. fig. 4.3, bottom), the results reported here are based on combustion experiments with H_2.

As already mentioned, fuel and air flows correlate according to the estimation of $\dot{m}_{fuel,theor}$ described in the previous section. Thus, the mixing quality should correlate with the data represented in figure 4.3. During the experimental runs, however, fuel and air flows are corrected with a term *RF* that stays for *real factor* and contributes balancing the effect of heat losses in the experimental setup:

$$\dot{m}_{fuel,real} = RF \cdot \dot{m}_{fuel,theor} \tag{4.4}$$

During operation, *RF* is adjusted in order to reach the desired reheating temperature of 1000 °C at least in one of the positions monitored in the chamber. The value of *RF* decreases as $\dot{V}_{inert}$ increases (see section A.1.4). For the experimental conditions tested, this means that the *RF* applied with $\dot{V}_{inert} = 10\,\mathrm{Nl/min}$ reaches values close to 4, whereas at $\dot{V}_{inert} = 40\,\mathrm{Nl/min}$ a *RF* of 1 can be used. This is reflected in the experimental results represented in figure 4.4.

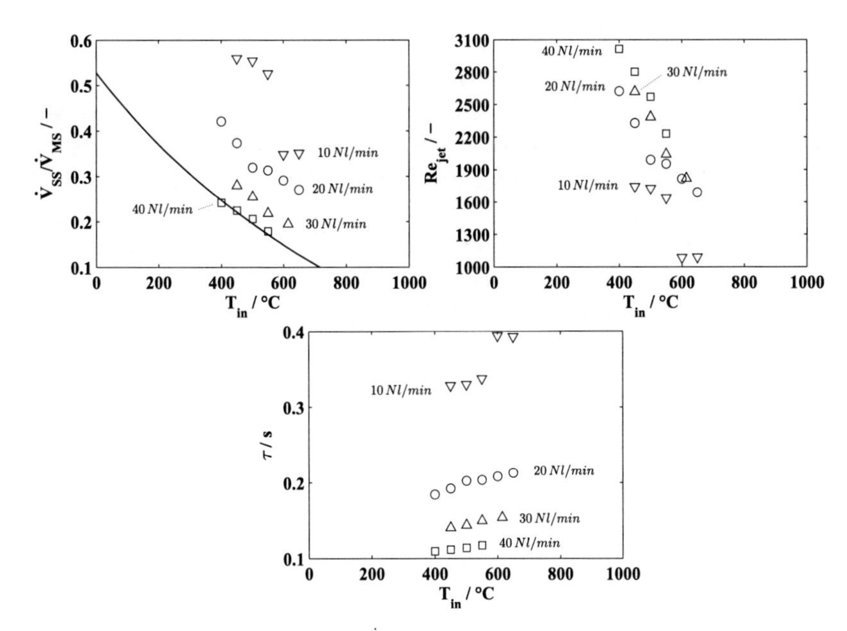

Figure 4.4: Measured and theoretical $\frac{\dot{V}_{SS}}{\dot{V}_{MS}}$ ratio (top, left), Re_{jet} (top, right) and residence time τ (bottom) for several $\dot{V}_{inert}$ against T_{in} and stoichiometric H_2 combustion.

The upper left representation shows that under experimental conditions $\frac{\dot{V}_{SS}}{\dot{V}_{MS}}$ increases as $\dot{V}_{inert}$ is reduced due to the effect of RF. As far as Re_{jet} is concerned, similar values are achieved as a function of the inlet temperature T_{in} for $20\,\mathrm{Nl/min} \geq \dot{V}_{inert} \geq 40\,\mathrm{Nl/min}$. However, as shown in the upper right representation, this is not the case for operation with $10\,\mathrm{Nl/min}$, where RF is not large enough to compensate the reduced volume flows. The lowest representation shows that the residence time of the gases in the chamber remains proportional to the inlet volume flow, regardless of the increasing RF values as the inlet volume flow is reduced. As discussed later on, this might have an effect on the fuel conversion attained during combustion.

Although the tracer experiments reported in section 3.3.2 suggest that Re_{jet} plays a secondary role in the mixing quality of the chamber (cf. fig. 3.8), the low values reached during operation with $\dot{V}_{inert} = 10\,\mathrm{Nl/min}$ seem to correlate with a reduced uniformity in the temperature distribution at low operation loads. Possible causes therefor are next discussed.

Temperature uniformity

Even though the lowest operation load $\dot{V}_{inert} = 10\,\mathrm{Nl/min}$ exhibits the largest $\frac{\dot{V}_{SS}}{\dot{V}_{MS}}$ ratio, temperature homogeneity based on the selected performance indicators show the largest deviations. This observation is supported by the data reported in figure 4.5 where ΔT_{bulk} is plotted against T_{in}. Besides reproducibility, data obtained during operation with $\dot{V}_{inert} = 20$ and $30\,\mathrm{Nl/min}$ shows the best temperature uniformity from all configurations tested. As shown in figure 4.5, b and c, measured ΔT_{bulk} lays in both cases in a well defined region between 40 and $80\,\mathrm{K}$. Data obtained with $\dot{V}_{inert} = 40\,\mathrm{Nl/min}$ also concentrates in a tight region, however beyond $\Delta T_{bulk} = 80\,\mathrm{K}$ and thus, characterized by worse homogeneity than in the former cases.

The measured ΔT_{bulk} for operation with $\dot{V}_{inert} = 10\,\mathrm{Nl/min}$ is very disperse ($20\ K < \Delta T_{bulk} < 130\ K$) and shows no apparent trend. An explanation therefor may rely in the fact that the fuel is injected through the outer wall of the chamber, perpendicular to the direction of the main flow. Since Re_{jet} exhibits the lowest values for operation with $\dot{V}_{inert} = 10\,\mathrm{Nl/min}$, the impulse of the air jet is presumably not large enough to entrain the fuel in a reproducible manner. Consequently, segregated reaction zones may establish in the bulk, causing the disparity in the measurements reported. Since visual assessment of the combustion process is not possible, this postulation can only be clarified with support of simulation results (cf. section 4.3).

Nevertheless, the observations based on the measurement of ΔT_{bulk} can be complemented

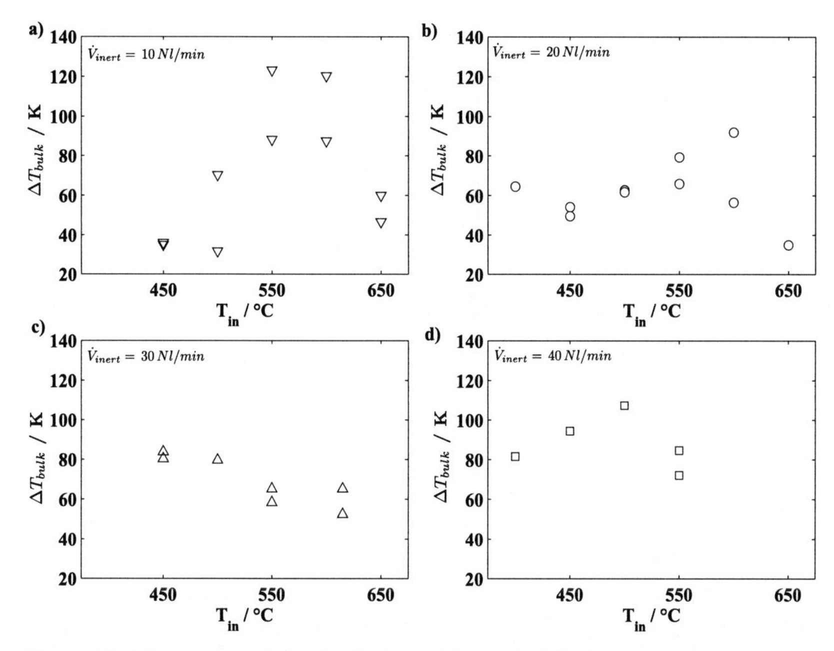

Figure 4.5: ΔT_{bulk} measured 30 s after ignition of the gas bulk for H_2 combustion against T_{in}. Plots a to c: $\dot{V}_{inert} = 10$, 20, 30 and 40 Nl/min. Data extracted from figure A.1.

with information about the maximal temperature difference in the lower chamber plate ΔT_{plate}, as summarized in figure 4.6. Temperature gradients in the lower plate can be explained by the localized incidence of the hot air jets, whereas the rest of the plate is cooled down by the effect of the inlet flow. This is in good agreement with the fact that the lowest gradients measured correspond to operation with $\dot{V}_{inert} = 10\,\text{Nl/min}$, which exhibits the lowest values of Re_{jet}. A combined interpretation of the two temperature uniformity indicators considered suggests that, at low operation loads, a non homogeneous reaction zone concentrates in the middle or upper region of the chamber. Accordingly, operation at larger operation loads should be striven in order to take advantage of improved homogeneity in the chamber.

As already asserted, larger inert volume flows between 20 and 30 Nl/min represent a robust operation range regarding temperature uniformity in the bulk of the chamber. This reproducibility is also given for the temperature gradients in the lower plate. Beyond this operation load, figure 4.6 reveals a general trend consisting of an increase in ΔT_{plate}. Again, this observation correlates with the behaviour already observed in figure 4.5 for operation with 40 Nl/min and suggests that an increase in the operation load does not further contribute improving the mixing quality of the chamber content.

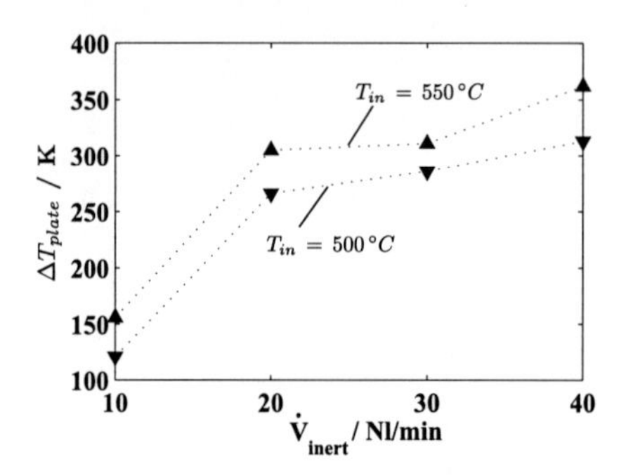

Figure 4.6: Largest ΔT_{plate} measured for $T_{in} = 500$ and 550°C against $\dot{V}_{inert}$. Data extracted from figure A.1.

A further consideration deriving from the information contained in figure 4.6 is the increase of the temperature inhomogeneity as the inlet temperature is increased. In such a case, the air and fuel flows are proportionally reduced, affecting the homogeneization process. Although experimental data with larger inlet temperatures is not available, this trend is not expected to extend over the whole range of T_{in}. The reason therefor relies in the fact that ΔT_{plate} directly correlates with the difference between the maximal temperature achieved T_{max} and the inlet temperature of the flow. Since the fuel and air flows are controlled in order to mantain $T_{max} = 1000$ °C, an increase in T_{in} implies a decrease in the temperature difference $T_{max} - T_{in}$. Accordingly, ΔT_{plate} must exhibit a maximum which, assuming a linear increment of the former with increasing T_{in}, is estimated to lay beneath 375 K. Although gradients in this order of magnitude have been measured several times during the experimental runs (cf. fig. 4.6), visual examination of the chamber internals after more than 80 h operation does not reveal any structural damage or deformation, corroborating the possibility to safely operate under conditions inducing such temperature gradients within the chamber.

Fuel conversion

Additionally to the evaluation of the chamber performance based on the temperature distribution, the current experimental setup was provided with the possibility to analyze the oxygen content in the bulk of the chamber during operation. Based on this information, fuel conversion achieved under several operating conditions can be calculated. The results for three different inlet temperatures and the inert volume flow ranging from 10 to 40 Nl/min are reported in figure 4.7.

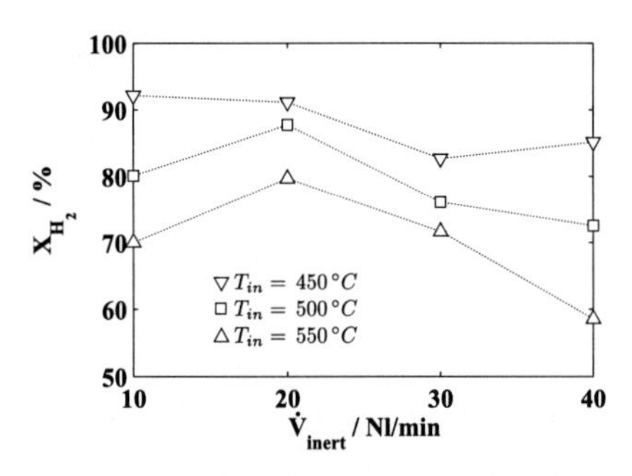

Figure 4.7: Hydrogen conversion in the combustion chamber as a function of T_{in} and $\dot{V}_{inert}$.

Provided that fuel and air flow rates and hence, the mixing quality in the chamber are inversely proportional to T_{in}, an increase in the latter correlates with a fuel conversion decaying trend. Equally, conversion dependency on $\dot{V}_{inert}$ is in agreement with the evaluation of temperature uniformity previously reported. The lowest (10 Nl/min) and largest (40 Nl/min) operation loads tested show the largest sensitivity to changes in T_{in}. This can be attributed to the large sensitivity of the mixing quality towards variations in the side-to-main flow ratio, in agreement to the poor temperature distribution homogeneity already discussed on the basis of figure 4.5. Despite thereof, operation with 10 Nl/min and the lowest T_{in} tested exhibit the largest conversion measured, which could be a positive effect of the large residence times at low operation loads (cf. fig. 4.4). The operation range between 20 and 30 Nl/min shows an almost constant sensitivity towards an increase in T_{in}. Operation at 20 Nl/min, however, exhibits larger conversions, presumably due to the somewhat larger residence times than during operation at 30 Nl/min.

Remarks on the optimal operation range

Based on the temperature uniformity and the fuel conversion reported, the best operation conditions within the analyzed T_{in} range tested can be achieved for an operation load of $20\,\mathrm{Nl/min} \leq \dot{V}_{inert} < 30\,\mathrm{Nl/min}$. The flow pattern established during operation with $\dot{V}_{inert} = 20\,\mathrm{Nl/min}$ exhibits not only good temperature homogeneity properties, but also the largest fuel conversions for the set of operation conditions tested.

The available results for operation with PSA off-gas reveal similar trends to those reported for H_2. As with the latter, operation beyond $\dot{V}_{inert} = 20\,\mathrm{Nl/min}$ does not exhibit any relevant enhancement of the mixing quality or the temperature uniformity since the use of PSA off-gas implies larger $\frac{\dot{V}_{SS}}{\dot{V}_{MS}}$ ratios and Re_{jet} than in the case of H_2.

It is worth noticing that the optimal operation range above mentioned is strictly related to the geometric dimensions of the combustion chamber studied so far. Provided that the same design has been applied in the validation of the regeneration strategy under reverse-flow operation (see chapter 5), this analysis directly affects the parameters set during the experimental runs. Accordingly, $\dot{V}_{inert}$ should be in the range between 10 and $20\,\mathrm{Nl/min}$, considering that the total inert volumetric flow increases in the direction of the flow through addition of fuel and combustion air in each of the chambers.

Simultaneously, the need to define a methodic approach to identify proper operation ranges for differing chamber designs arises. This issue can be tackled making use of simulation tools and is reported in the following section.

4.3 Systematic combustion chamber design based on CFD

The results reported in the previous section verify the applicability of the combustion chamber concept for the stoichiometric regeneration strategy. Furthermore, they have contributed to identify a proper operation range, for which the advantages of the chamber can be best exploited. However, the analysis does not enable a systematic design of such a chamber for operation loads strongly differing from those targeted in this work. Accordingly, the task of the current section is to analyze the possibility to use the experience gained in order to establish a preliminary design method based on CFD simulations.

Based on the available experimental data base, simulation results can be validated based on experimental observations. Relevant performance indicators for the combustion chamber have to be derived and the limits of the available simulation tools to design such a system in an accurate but also efficient way need to be defined. Main target is thus, to find a compromise between model complexity and computational expensiveness for future chamber design or optimization.

The strategy followed, which is extensively described by Speidel in [97], consists in the simulation of the current chamber using a detailed, multicomponent model including chemical reaction. In order to limit the complexity of the system, only combustion of pure hydrogen has been considered and modelled using the so called Eddy Dissipation Model (EDM), which builds on the assumption that reaction is only limited by the molecular mixing of the reactants involved. Based on the results obtained, mixing quality and reaction performance are characterized. A subsequent simplification of the mathematical model is then performed by reduction of the former to describe a multicomponent, nonreactive -and hence isothermal- system and, in the simplest case, a single-component, isothermal model. Comparison of the results obtained with the three models enables to determine the degree of simplification allowed, in order to obtain relevant information characterizing the behaviour and performance of the chamber.

The combustion chamber has been modelled and simulated with the software ANSYS CFX® [98]. The results discussed correspond to stationary simulations of the chamber, which has been discretized using a computational grid containing 850.000 cells. Additional 350.000 elements are used to discretize the packing sections below and above the chamber. Furthermore, an upwind scheme has been used to approximate the convective terms. In all cases, turbulence is described with the RNG(Renormalized Group)-k-ε-model. This configuration enables a robust and less computationally expensive analysis of the chamber in comparison to reference simulations performed with a finer grid (up to 4 million cells) and a high-resolution approximation scheme. Moreover, the trends described by the performance indicators defined to evaluate the results (see following section 4.3.1) can be properly reproduced regardless of the lower accuracy of the results.

4.3.1 Performance indicators

As already discussed, optimal operation of the chamber is characterized by a homogeneous temperature distribution and preferably, large fuel conversions. These two parameters correlate directly with the homogeneity achieved during the mixing process. Hence, the performance indicators used for the current analysis need to be conclusive regarding the mixing quality of the gases in the chamber.

Recirculation rate (K_V)

As already described in section 3.1, recirculation rate K_V and temperature are the main parameters that enable defining the regime at which a conventional FLOX® combustor operates, namely with flame -stable or unstable- or under flameless conditions. The definition is given by equation 3.1 and literally describes the amount of recirculated off gases in relation to the *fresh* fuel and combustion air fed to the combustion chamber at a given point. This definition, however, does not provide any explicit information about the flow characteristics in the chamber and assumes that homogeneous mixing of the *fresh* inlet with the combustion off gases occurs. Thus, K_V can be interpreted as a measure of the dilution degree of the reactive gases fed to the chamber.

It was task of this study to determine, whether the correlation between temperature and K_V deliver similar information about the combustion regime taking place in the chamber under study. Speidel visually summarizes in figure 4.8 the meaning of the recirculation rate as defined in the current study.

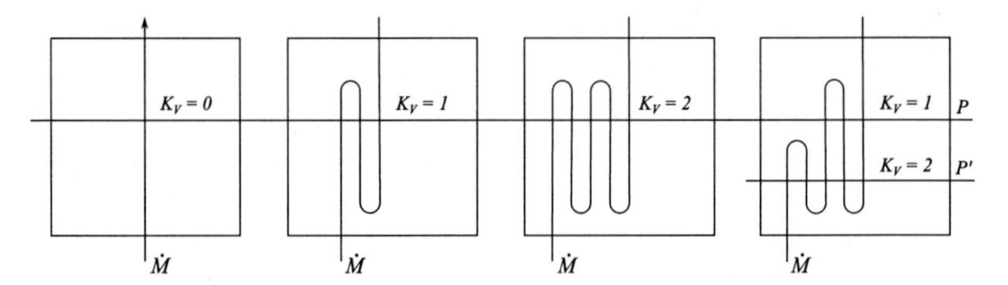

Figure 4.8: Visual description of the recirculation rate K_V of the mass flow $\dot{M}$ in a plane P (or P') within the combustion chamber [97].

K_V describes the relation between the recirculating and the main inlet/outlet streams and is dependent on the chamber region at which it is evaluated. In order to account for this effect, K_V has been defined at several planes representing cross sections perpendicular to the main flow direction (from bottom to top), distributed along the height of the chamber. A total of 26 equidistant planes have been used in the current study. The total recirculated mass flowing through a plane P is defined as follows

$$\dot{M}_{Rec,P} = \frac{1}{2}\left[\left(\sum_{P}|\dot{M}_{Z,P}|\right) - \left(\dot{M}_{inert} + \dot{M}_{fuel} + \dot{M}_{air}\right)\right] \tag{4.5}$$

The previous definition of $\dot{M}_{Rec,P}$ assumes that the inert, fuel and air mass flows cross at least once each plane in the chamber before leaving it. Thus, they must be substracted from the summation of the absolute value of the mass flows in each of the volume cells Z contained in a plane P. The factor $1/2$ derives from the fact, that each recirculating flow crosses the same plane at least twice. The characteristic indicator chosen to describe the recirculation ratio in the combustion chamber is defined as the maximal estimated K_V

$$K_{V,max} = max\left(\frac{\dot{M}_{Rec,P}}{\dot{M}_{inert} + \dot{M}_{fuel} + \dot{M}_{air}}\right) \tag{4.6}$$

If the multicomponent model is used, the same definition can be applied to describe the recirculation of each of the substances present in the chamber:

$$K_{V,i}(P) = \frac{\frac{1}{2}\left[\left(\sum_P |\dot{M}_{i,Z,P}|\right) - \dot{M}_i^{fresh}\right]}{\dot{M}_i^{fresh}} \tag{4.7}$$

Mixing quality of component i (Mix_i)

Contrary to conventional FLOX® combustors, homogeneous mixing of all gases in the current system cannot be taken for granted. This is of particular importance provided that the operation concept based on stoichiometric combustion implies an external fuel supply. Thus, reaction performance is strongly dependent on the homogeneous distribution of all reactants within the chamber volume, which in turn, can be characterized making use of the degree of mixing of each component.

Since the recirculation is generated by the air injected, it can be assumed that large recirculation rates correlate with large air mixing qualities. This reciprocity is not implicitely given for the mixing quality of the fuel, depending on the position and the type of the side feed, as well as on the fuel flow rate. In order to characterize this dependence, the mixing quality has been defined as follows:

$$Mix_i = 1 - \frac{\sigma_i}{\sigma_{i,max}} \qquad 0 \leq Mix_i \leq 1 \tag{4.8}$$

with σ_i representing the average deviation of the mass fraction of the component i from that of an ideal, homogeneous mixture

$$\sigma_i = \frac{1}{M_{tot}} \cdot \sum_V |\overline{w_i} - w_{i,Z}| \cdot \rho_Z \cdot V_Z \tag{4.9}$$

The mass fraction of component i is averaged in reference to the total mass of the chamber volume, considering the mass of i contained in each discretization cell

$$\overline{w_i} = \frac{1}{M_{tot}} \cdot \sum_V w_{i,Z} \cdot \rho_Z \cdot V_Z = \frac{1}{\sum_V \rho_Z \cdot V_Z} \cdot \sum_V w_{i,Z} \cdot \rho_Z \cdot V_Z \tag{4.10}$$

With this definition of Mix_i, it is possible to accurately estimate the quality of mixing of the fuel, even though the latter exhibits a considerably lower average mass fraction than the rest of the components present in the chamber [97].

4.3.2 Simulation results - Correlation with performance indicators

The majority of the results discussed in the current section correspond to operation under stoichiometric conditions. Thus, the air and fuel flow rates are subject to the dependencies from $\dot{V}_{inert}$ and T_{in} defined in section 4.2.1. Furthermore, it is worth noticing that all simulations have been performed using hydrogen as fuel. The effect of using alternative fuels on the chamber performance is solely taken into account from the perspective regarding the flow distribution and mixing quality. In the current case, PSA off-gas is considered. The latter exhibits a slightly lower heating value than hydrogen on a molar basis. Additionally, due to the large difference in the molar weight of both fuels $\left(\frac{MW_{PSA}}{MW_{H_2}} \approx 10\right)$, the momentum of the fuel flow is considerably larger in case of using PSA off-gas instead of hydrogen for a given heating value. In order to account for this effect, emulation of PSA off-gas is performed by supply of the corresponding hydrogen amount to match the heating value and correcting the impulse of the flow by adding an inert component (N_2).

In order to illustrate the different operation regimes, two configurations for the fuel supply are considered:

- fuel is diluted and fed together with the inert main flow $\dot{V}_{inert}$ resembling operation conditions reported in chapter 3, and
- fuel is fed to the chamber through an extra side-feed, corresponding to the current stoichiometric combustion approach.

In both cases, combustion air is supplied through the injection nozzles on the top of the chamber and the inert main stream (*MS*) flows through the chamber from bottom to top. Figure 4.9 depicts some of the possible configurations. Thereby stays *SF* for side-feed, 1 and 2 for different positions within the perimeter of the chamber, and *M* and *B* for **m**iddle and **b**ottom respectively, referring to positions within the chamber height.

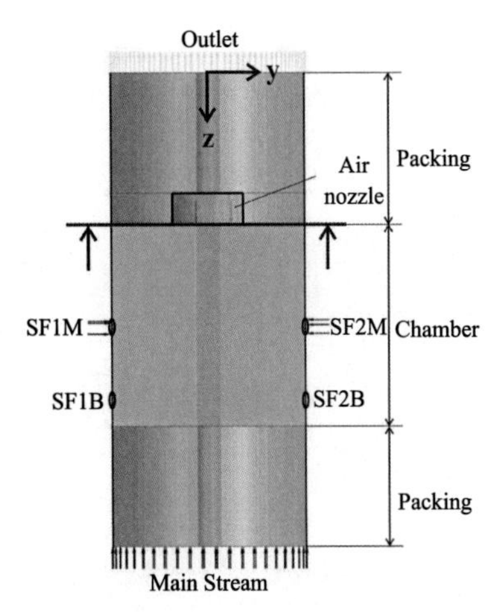

Figure 4.9: Sketch of the combustion chamber modeled in ANSYS CFX. SF1U, SF1B, SF2M and SF2B define several side-feed configurations for external fuel supply to the chamber.

Parameters influencing flameless operation

As already advanced in section 4.3.1, the definition of K_V in the current study has a different implication than the one used to characterize the operation regime of a conventional FLOX® burner (see section 3.1). It was therefore task of the current analysis to determine the adequacy of K_V and Mix_i in order to characterize flameless operation.

The feasibility to resolve flame formation in the chamber has been validated through computation of several experimental configurations exhibiting flame formation. The formation of flame fronts is properly reproduced and can be qualitatively detected in the simulated temperature distributions. Thus, taking the performance indicators under such conditions as reference and analyzing the sensitivity of the operation regimes towards variations of the latter, it is possible to determine the effects playing a major role in the suppression of flame formation.

Figure 4.10 represents an example of this procedure, which shows the temperature distribution within a frontal section cut of the chamber as sketched in the leftmost representation. In both cases considered, fuel is supplied diluted in the main flow. Heat released in both cases is 0.5 kW and the air flow rate is kept constant. However, the representation on the left corresponds to understoichiometric combustion ($\lambda = 0.32$), whereas the right one corresponds to combustion with $\lambda = 1$ (i.e. the fuel amount in the main flow has been reduced and the inert flow rate accordingly increased). Whereas the former clearly exhibits the formation of flame regions, the latter shows a far more homogeneous temperature profile.

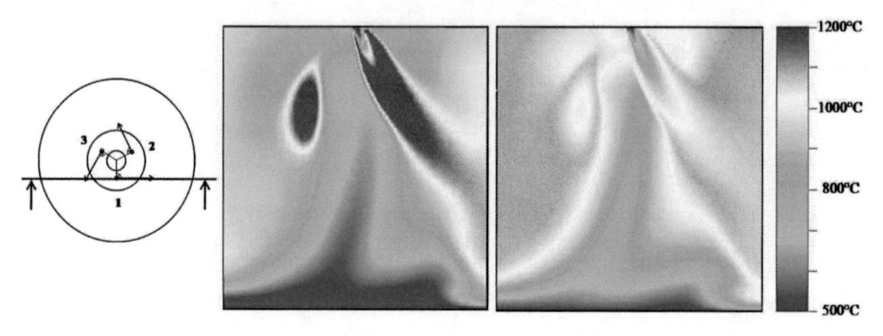

Figure 4.10: Temperature distribution in a frontal section of the chamber during combustion with 0.5 kW heat release, fuel premixed in the main flow and constant air flow. Left: combustion with large fuel excess ($\lambda = 0.32$); right: combustion under stoichiometric conditions ($\lambda = 1$) [97]. Detailed information about the parameter settings for simulation is contained in section A.1.3.

Provided that the main responsible for the recirculation rate is the effect of the air jet, it becomes clear that both configurations must exhibit similar K_V values, which lay close to 1.3. Despite being far below the minimal K_V of $2-3$ proposed for flameless operation in conventional FLOX® burners, no flame formation can be detected in the second configuration. This can be attributed to the definition differences between K_V in section 3.1 and $K_{V,max}$ in section 4.3.1. In the former case, K_V represents a measure of the dilution degree of reacting and combustion off-gases under the assumption that homogeneous mixing of the chamber content is given. In the current case, however, the dilution effect achieved through a large mixing degree is not implicitely contained in the definition of the recirculation ratio (cf. fig. 4.8). Thus, the results shown in figure 4.10 suggest that dilution of the reacting gases with inert components plays a major role in the suppression of flame regions and in the establishment of a homogeneous temperature distribution within the chamber.

In case that fuel is diluted in the main flow, stoichiometric operation represents a great advantage in this respect, since the fuel-to-inert ratio, defined as $\frac{\dot{V}_{fuel}}{\dot{V}_{inert}}$ and serving as indicator of the fuel dilution, is clearly reduced. However, this behaviour is not necessarily given if the fuel is independently supplied to the combustion chamber, as in the current reheating strategy. In such a case, the homogeneous distribution of the fuel within the chamber, described by the performance indicator Mix_{fuel} plays a major role.

In order to study the correlation between Mix_{fuel} and the formation of flames, an operation configuration with fuel supply in the main flow has been taken as reference and compared to simulation results obtained with similar operation parameters but separated fuel supply. The different configurations tested are summarized in table 4.1. MS stays for fuel supply diluted with the main stream and $SF1M$ and $SF2M$ for an external side-feed according to the sketch represented in figure 4.9. According to the performance indicators estimated for the different cases and reported in the last two columns of table 4.1, the largest effect when varying the position of the fuel supply is reflected in the mixing quality of the latter, whereas the value of the recirculation ratio remains within a narrow range.

Case	**Fuel supply**	$\dot{V}_{MS}/Nl/min$	$T^{in}/°\mathrm{C}$	$RF/-$	$K_{V,max}/-$	$Mix_{fuel}/-$
1	MS	30	500	1,3	1,36	0,97
2	SF1M	30	500	1,3	1,48	0,71
3	SF2M	30	500	1,3	1,54	0,66

Table 4.1: Simulation parameters and performance indicators of a combustion chamber run at a constant operation point with different positions for the fuel supply (cf. fig. 4.9).

The consequence of the lower fuel mixing quality becomes evident in figure 4.11. From top to bottom, the temperature profiles for the three cases summarized in table 4.1 are shown. From left to right, the temperature distribution in the transversal section at the chamber outlet as well as cuts of the frontal section in intervals of $30°$ are shown.

Following the expectations, fuel supply together with the main stream exhibits the most homogeneous temperature distribution (figure 4.11, top), correlating with the large mixing quality of the fuel ($Mix_{fuel} = 0.97$). On the contrary, a variation of the position of the fuel feed is accompanied by a decrease in the mixing quality and thus, in the temperature homogeneity (figure 4.11, middle and bottom). Lower mixing qualities occur for those configurations in which the fuel flow is not properly entrained by the air jets, resulting in a portion of the former that leaves the chamber without having reacted. This is accompanied by a reduction in fuel conversion and the development of temperature inhomogeneities that occur in those regions where air and fuel encounter (cf. temperature uniformity discussion in section 4.2.2). Temperature maxima resembling typical flame fronts may occur, reaching

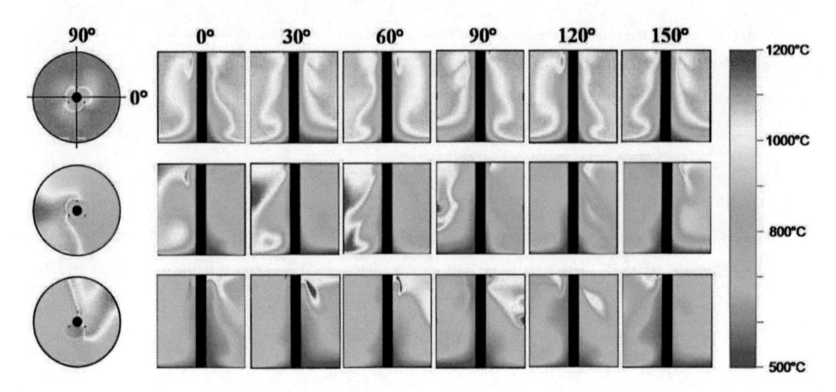

Figure 4.11: From left to right: temperature profiles at the outlet and along the vertical axis of the chamber. From top to bottom: operation points corresponding to cases 1 to 3 reported in table 4.1 [97].

estimated values between 1200 and 1400 °C in the last two cases considered respectively. Fortunately, and in agreement with the experimental results so far discussed, critical temperatures are primarily reached at the outlet of the air nozzles, so that no material damage is to be expected.

Nonetheless, this analysis provides valuable information to be considered in future designs of similar combustion chambers. The strong dilution of the fuel when operating under stoichiometric conditions enables flameless operation even at recirculation rates K_V between 1 and 2. Moreover, the mixing quality of the fuel directly affects its conversion and consequently, temperature homogeneity. Dilution of the fuel in the main stream provides the largest mixing qualities. In a system where several chambers are to be simultaneously supplied with fuel, the fuel should be introduced together with the main flow. Intuitively, distribution of the fuel along the perimeter of the chamber bottom, using appropriate geometries to boost the momentum of the fuel when entering the chamber, represent an alternative to improve its homogeneization.

4.3.3 Reduced models for the combustion chamber design

According to the results discussed in the previous section, large fuel mixing qualities Mix_{fuel} are the determining factor promoting operation under flameless conditions, since the recirculation ratio K_V plays a secondary role in the dilution of the fuel and thus, in the supression of flame formation. Therefore, the main aspect to be adressed in this section is, to which extent the mathematical model used can be simplified without suffering a relevant information loss. Again, the model reduction is considered for the two fuel supply configurations already mentioned.

Fuel diluted in the main stream

The first simplification step consists of neglecting the chemical reaction and considering isothermal operation of the chamber. It can be generally stated, that the estimation of the recirculation rate K_V and the mixing quality of the different components Mix_i slightly overestimate the results obtained with a reactive system (s. fig. 4.12). However, this difference arising from temperature gradients and their influence on the physical properties of the fluid, does not affect the trend of the mixing qualities and in particular, that of the fuel.

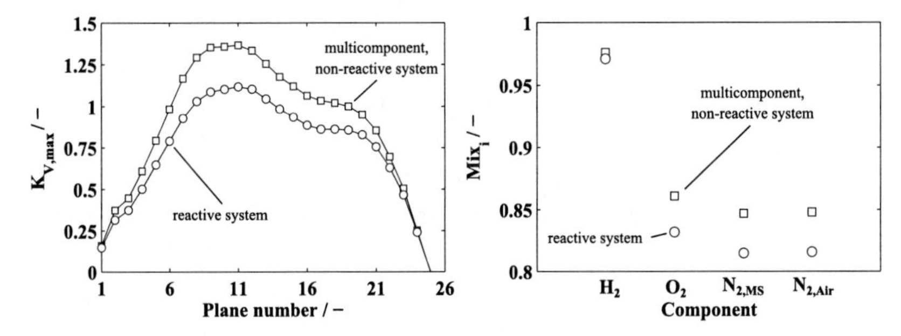

Figure 4.12: Comparison of the K_V distribution along the chamber (left) and the mixing quality Mix_i of several components in it for simulations with and without chemical reaction (right). Fuel diluted in the main flow [97]. Plane number stays for each cross section along the chamber height, at which the recirculation ratio is evaluated (s. sec. 4.3.1).
Operation parameters: $\dot{V}_{MS} = 30\,\mathrm{Nl/min}$; $\frac{\dot{V}_{SS}}{\dot{V}_{MS}} \approx 0.2$; operation load 0.42 kW.

Simulation data available for the configuration in which fuel is fed together with the main flow reveals a direct proportionality between $\frac{\dot{V}_{SS}}{\dot{V}_{MS}}$ and $K_{V,max}$, corroborating the observations already reported based on experimental results. Furthermore, for the same data set, $K_{V,max}$ and Mix_i also exhibit a proportionality trend. Hence, for $K_{V,max}$ in the range between 0.8 and 1.8, large fuel mixing qualities and thus fuel conversions between $95-99\%$ can be expected. Based on this observation, it can be concluded that the estimation of a single parameter, in this case $K_{V,max}$, under isothermal conditions is sufficient in order to correctly characterize the performance of the chamber when the fuel is pre-diluted in the main stream. Moreover, it has been proven that an isothermal, single component model provides reliable estimations of the recirculation ratios and is consequently an appropriate approach to perform exploratory or optimization simulations with a reduced computational effort.

Fuel supplied through a side feed

In contrast to the previous case, correlation of $K_{V,max}$ with mixing quality indicators in the chamber is not implicitely given for those configurations where the fuel is fed through a side-feed. In such a case, the fuel mixing degree represents the most important indicator, directly correlating with the conversion rates and thus, with the temperature homogeneity in the chamber.

Both the position of the fuel feed and the momentum of the fuel flow, play an important role in its mixing degree. An accurate estimation of the latter is necessary in order to properly characterize the operation performance of the chamber. For this purpose an isothermal, multicomponent model can be used. Further simplification can be achieved by considering a single component model. However, the latter requires the introduction of tracer particles in order to evaluate the mixing behaviour.

When performing simulations with a single-component model it is advisable to use the properties of the main substance in the system, such as nitrogen in the current case. Thus, if the physical properties of the fuel used significantly differ from those of the single component used (as it is the case when analyzing hydrogen combustion), the validity of the results is not necessarily given. An accurate estimation is only granted when both the fuel mass flow rate and the momentum of the flow are adjusted to those in the real system. This proceeding generally involves grid modifications to adapt the geometry of the fuel inlet, which distorts the computational efficiency that might be gained by using the simplified model. Thus, a single-component model is only advisable for preliminary estimations and if the properties of the fuel and the single component used for simulation are similar enough, so that matching the mass flow suffices for a first evaluation.

Based on the simulation results, the relevant trends in the performance of the chamber operated with a fuel side-feed can be described. For the model system considering hydrogen as fuel, significantly lower Mix_{fuel} values, between 40 and 80%, than for comparable operation points with fuel mixed in the main flow have been observed. This strongly affects fuel conversion and consequently, temperature homogeneity in the chamber. Recirculation $K_{V,max}$ and Mix_i are almost independent from the main flow, while the momentum of the fuel flow is the parameter showing the largest sensitivity. Thus, those operation ranges favouring larger fuel mass flow rates improve the performance of the chamber.

Operation with PSA off-gas as fuel presents several advantages in comparison to hydrogen. Since the heating value of the former per unit mass is significantly lower than that of hydrogen, the momentum of the fuel flow for an equivalent heating power is almost ten times larger in the first case. Thus, even though similar $K_{V,max}$ values are estimated for both fuels, the Mix_{fuel} values are significantly larger for the case with PSA off-gas. Using the latter, improvements of 25% in the conversion achieved can be expected. Furthermore, the significant presence of inert components in the fuel itself (i.e. CO_2), contribute to suppress flame formation due to their diluting effect.

4.4 Summary

The experimental results reported in this chapter verify the adequacy of the stoichiometric combustion strategy to overcome the limitations of the initial reheating strategy reported in chapter 3, and in particular the effects of undesired after reaction during combustion under fuel rich conditions. Furthermore, a detailed analysis of the chamber performance based on the evaluation of temperature homogeneity and fuel conversion during operation enables the definition of an optimal operation range for a combustion chamber with the geometry and the design of choice for the current study.

CFD simulations have proven to qualitatively reproduce those characteristics observed experimentally, such as flame formation. Furthermore, the systematic evaluation of previously defined performance indicators and their correlation with the different operation regimes obtained, enable the identification of relevant parameters affecting homogeneity of operation.

For operation under stoichiometric conditions and consequently, with separate air, fuel and inert supply, recirculation turns to be a necessary but not sufficient condition to guarantee homogeneity in the chamber. Mixing quality of the different components represents the determining factor and suggests that, for the case under study, room for improvement, primarily affecting the introduction of fuel into the chamber, is given.

Chapter 5

Regeneration strategy under reverse-flow operation

Main goal of the experimental analysis reported in this chapter is to verify the suitability of the regeneration strategy to efficiently reheat the RFR under periodic operation and in particular, the possibility to reheat the packed bed at a high temperature level avoiding excess temperatures.

5.1 Experimental setup

Figure 5.1 shows the process flow diagram of the experimental setup. The latter represents an extension of the setups previously used to analyze the behavior of a single combustion chamber and allows operation of a RFR under periodic operation. The different sections of the setup are next described.

Reactants supply: During reverse-flow operation, the reactor feed composition is periodically switched according to the requirements of the production and regeneration steps. The flow direction is simultaneously reversed actuating four pneumatic valves (valves $V1$ to $V4$). During production, methane as main constituent of natural gas (NG) and steam are fed in a steam-to-carbon molar ratio of $S : C = 3$. Therefore, steam is generated in situ from deionized, degased water. The latter is supplied with a reciprocating pump and is evaporated in a low-pulsation capillar evaporator developed at the ICVT with a capacity of up to $4\,\mathrm{kg/h}$ steam. Although steam is only required during the production period, it is continuously generated and by-passed during the regeneration step, in order to minimize dead times when performing flow reversal (valves $V5$ and $V6$). All other gaseous flows are adjusted using conventional mass flow controllers.

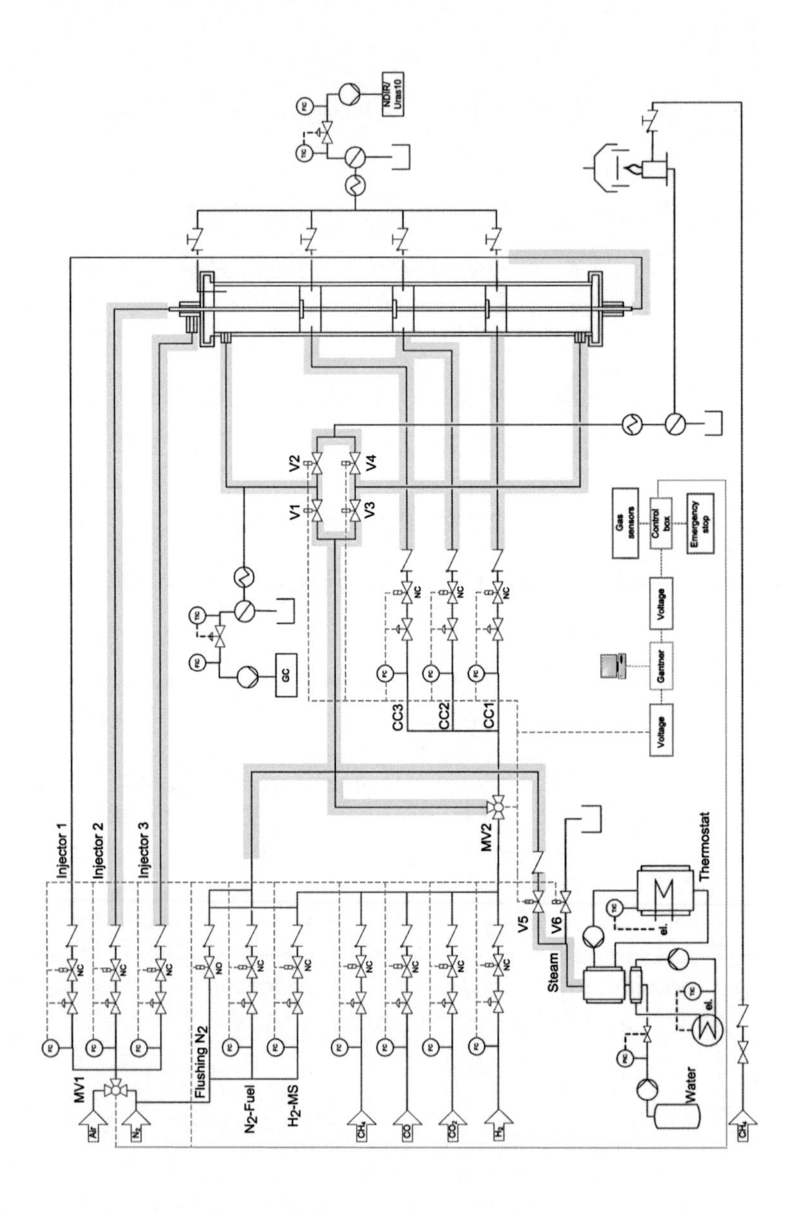

Figure 5.1: Process flow diagram of the RFR experimental setup.

During the regeneration period, the flow direction is reversed and the reactor is fed with an inert gas (N_2). Simultaneously, each of the combustion chambers embedded in the packed bed are fed with fuel and combustion air according to the considerations discussed in section 4.2.1. Operation both with H_2 or PSA off-gas as fuel is possible. The former is generally used during the start-up procedure, given its low catalytic ignition temperature. Once the packed bed has been preheated up to the desired temperature level, any fuel can be used to attain periodic operation. The use of PSA off-gas, however, corroborates the possibility to integrate the RFR in a decentral hydrogen production facility, as described in appendix E. Reactor piping is heated at temperatures between 150 and 200 °C amongst others, to avoid water condensation if steam is present.

Reactor: The RFR is the core item of the experimental setup. It consists of a packed bed with three embedded void sections that act as combustion chambers. The packing is structured in catalytic and inert sections. The first are packed with shell catalyst pellets of ceramic $\gamma - Al_2O_3$ carrier and a noble metal coating, with a particle diameter d_p averaging 2 mm. The catalyst shows activity for both reforming and combustion reactions and is supplied by courtesy of Umicore AG.

The arrangement and dimensions of the inert, active and combustion zones play a fundamental role in the behaviour of the RFR under periodic operation. For the current study, the configuration reported by Glöckler et al. in [41, 62, 99] is taken as reference and the inert combustion sections are replaced by flameless combustion chambers. A representation of the reactor tested is shown in figure 5.2, whereas the detailed drawings of the construction can be seen in appendix B.3. The reactor dimensions and the construction materials are summarized in table 5.1.

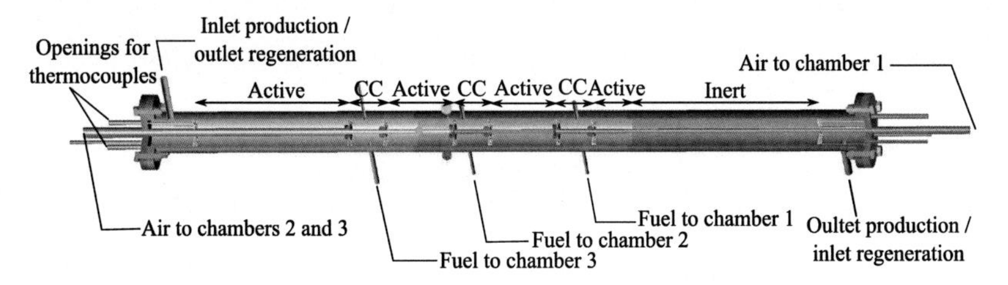

Figure 5.2: 3D sketch of the RFR in the experimental setup.

Dimension	Value	Units
Length (L)	900.0	mm
Active length	505.0	mm
Inert length	245.0	mm
Diameter	50.0	mm
Wall thickness	1.0	mm
Material	AVESTA 253MA	

Table 5.1: Dimensions of the tubular packed-bed reactor.

Each combustion chamber is provided with a fuel and air supply as shown in figure 5.3. The fuel is supplied as a side stream placed at half height of the combustion chamber, whereas the air is injected through the nozzles placed on the top of the chambers. Air is supplied through concentric ducts positioned in the reactor axis. The details of the reactor construction and the proceeding followed for its assembly are described in detail in [100]. An extract thereof can be found in appendix B.

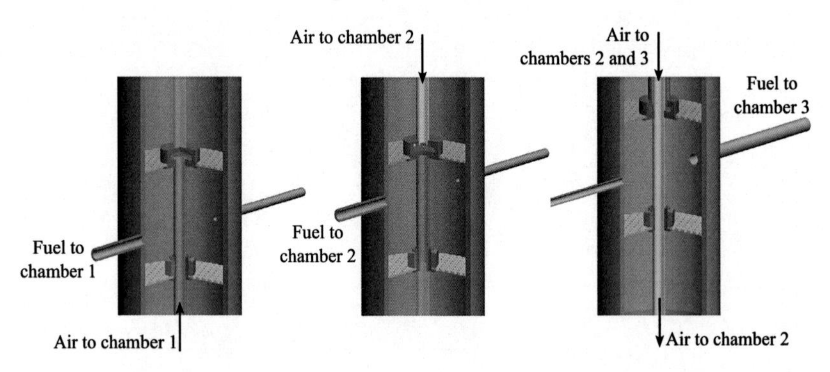

Figure 5.3: 3D sketch of the three combustion chambers mounted in the RFR (cf. figure 5.2).

The reactor ends are provided with flanges formerly sealed with Viton o-ring seals. However, in the course of the experimental operation, the flanges have been welded in order to avoid the risk of leaks in case of exceeding temperatures beyond those appropriate for the polymeric seelings (s. figs. 5.4 and B.7).

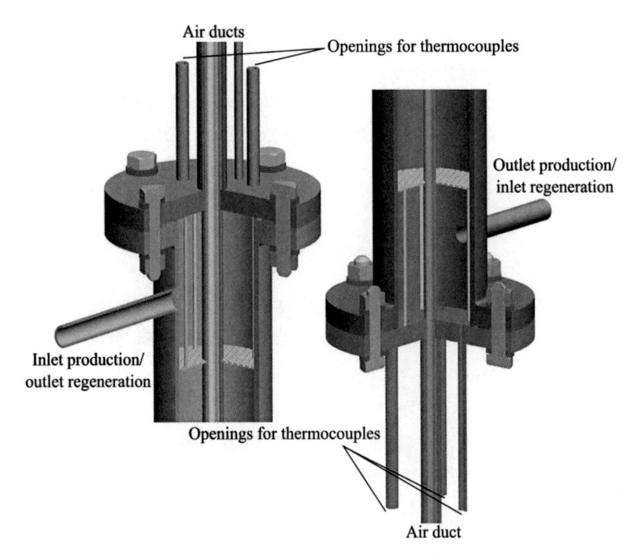

Figure 5.4: 3D sketch of both reactor ends (cf. figure 5.2).

Product disposal: The reactor effluents are cooled down to 10 °C for water condensation and sent to an atmospheric flare in order to combust any flammable components contained. The flare is provided with an auxiliary CH_4 flame, since during the regeneration step inert, non-flammable combustion off-gases leave the reactor and combustion in the flare would extinct otherwise. Monitoring of the flame activity is a relevant element of the safety strategy implemented in the automation system [100].

Analytics: The proces behaviour during reverse-flow operation is monitored on-line with a non-dispersive infrared spectrometer (NDIR-Photometer - Hartmann & Braun (ABB), Uras 10P) that detects CH_4, CO_2 and CO up to volume concentrations of 20%. After removal of the water contained, the dry gas sample is taken from the flow entering the flare with help of a membrane pump and a rotameter installed downstream, in order to guarantee the pressure level and the volume flow required to achieve reproducibility in the measurements (below 400 mbar; 20 – 100 l/h). The sampling system exhibits a delay of 20 to 25 seconds in the concentration measurements [1].

[1] Estimation based on comparison of FTIR and IR measurements. The former is fed with a sample directly taken from the reactor outlet and prior to the quench. An integrated sampling pump in the FTIR grants a continuous measurement unaffected by pressure changes in the system.

The effluent during the production period contains significant amounts of H_2O, H_2, CO_2 and CO and ideally none or only small amounts of CH_4, depending on the conversion achieved. After water condensation, monitoring of the last three substances allows an accurate evaluation of the reaction performance. During the regeneration period, only combustion products as well as N_2 are contained in the product flow. For PSA off-gas, monitoring of CO and CH_4 concentrations enables to determine whether full oxidation of the fuel introduced is achieved.

Automation and data acquisition: The complexitiy of the experimental setup requires an appropriate automation system that enables to coordinate the automatic operation of the different actuators in the setup, manage data acquisition, supervise the status of the plant and if required, activate safety measures based on predefined routines.

The system chosen consists of a programmable automation controller (PAC) of Gantner Instruments Test & Measurement GmbH, which synchronizes the controller functionality with the data transmission between the experimental setup and the host PC. For this purpose, several I/O's modules (e.bloxx) can be connected at each of the four RS 485 slave interfaces that are provided by the PAC. The latter communicates with the host PC via an ethernet interface [101].

As interface for the communication with the experimental setup, LabView from National Instruments is used. It allows to create custom-made graphical user interfaces for a comfortable operation and supervision of the experimental setup. For a detailed description of the structure of the routines used, the reader is referred to [76, 100].

5.2 Experimental procedure

Each experimental run consists of two different phases. During start-up, the reactor is heated up in order to establish the desired initial temperature profile. The latter should exhibit the characteristics described in section 2.3.1 and represented in figure 2.2, i.e. low temperature level at the reactor boundaries and an extended, high temperature plateau in the reactor center. As soon as this profile has been established, the second phase consisting of periodic operation can be initiated. Relevant details regarding the two experimental stages are next summarized.

5.2.1 Start-up procedure

During the start-up procedure, which extends over approximately 1 h, H_2 is used as fuel since it exhibits the lowest catalytic ignition temperature. Contrary to periodic operation, which alternates between production and regeneration, combustion conditions are established for both flow directions during start-up. The reactor ends and the piping are preheated up to $120 - 150\,°C$. The main flow, consisting of N_2, is periodically reversed every 300 s, while the flow rate is incremented from around $2\,Nl/min$ at the beginning of the stage, up to $10 - 15\,Nl/min$. Fuel and air flows are supplied according to the description in section 4.2.1. Figure 5.5 shows a characteristic temperature profile at the end of the start-up procedure, where the potential of the reheating strategy to establish a flat temperature plateau in the catalyst bed with cold ends at both entrance/exit sides can be identified already.

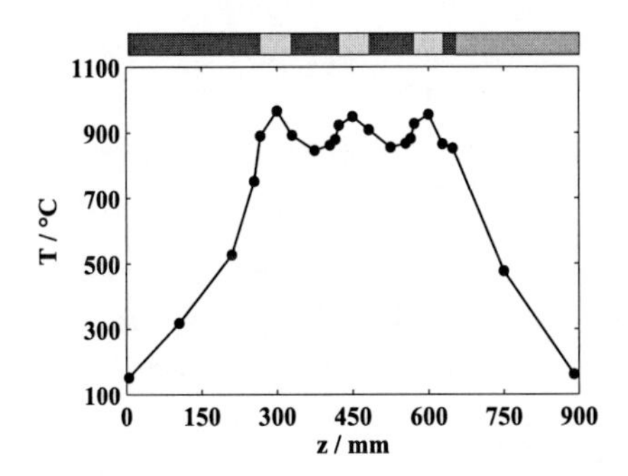

Figure 5.5: Initial temperature profile along the reactor length after the start-up procedure.

5.2.2 Periodic operation

Parameters for periodic operation are estimated prior to each experimental run. They are based on several process constraints, summarized schematically in figure 5.6. The operation parameters calculated preliminary are used for guidance and may be tuned during operation by hand.

Parameter estimation

The operation load during production is defined by the hydrogen yield, which in turn, is represented in terms of the continuous thermal power produced. For each operation load, the reactant flows during the production period are given and thus, the average thermal front velocity $w_{therm,prod}$ can be estimated (cf. eq. (2.10)). The amount of energy consumed by the reforming reaction Q_{prod}, as well as the distance travelled by the thermal wave can be evaluated for a given period length τ_{prod}.

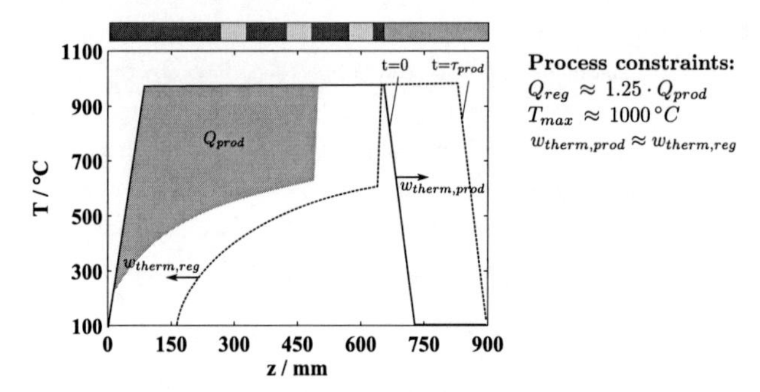

Figure 5.6: Simplified representation of the procedure to estimate operation parameters for periodic RFR operation.

The estimation of the operation parameters during the regeneration period is made under the assumption that $\tau_{prod} = \tau_{reg}$. The energy consumed by reaction needs to be reintroduced in the packed bed during regeneration. Conservative thermal efficiencies of $75-80\%$ to account for heat losses are assumed, so that the required energy supply is required to be $Q_{reg} \approx 1.25 \cdot Q_{prod}$. Since the fuel composition is given and full combustion of all flammable species is assumed, the fuel and combustion air flows over the regeneration period can be directly derived from Q_{reg}.

The minimal required amount of inert gas fed to the reactor during the regeneration period is first estimated, so that the average thermal velocity of the flow composed by the combustion products, air nitrogen and inert gas fulfills $w_{therm,reg} = w_{therm,prod}$. This first estimation is subject to a considerable uncertainty, since the fuel and air flow rate adjustment during regeneration implies a time and position dependant $w_{therm}(t,z)$, wich can not be easily averaged. Some representative operation parameters for operation with hydrogen as fuel are summarized in table A.5.

After start-up, the inert gas flow rate supplied during regeneration as well as the duration of the production and regeneration periods are two relevant parameters that can be controled in a simple way. Sensitivity of the operation towards variations in the latter parameters is discussed in section 5.3.

Cyclic stationarity

Operation stationarity is achieved when the same temperature profile is obtained at the end of two consecutive cycles, i.e. after two production or two regeneration periods. Despite manual tuning of the parameters during operation, stable operation states can be reached with maximal temperature deviations of around 3%. Figure 5.7 shows the evolution of the averaged and maximal temperature differences between two consecutive profiles at the end of production (upper figures) and regeneration (lower figures). Within 6 to 8 complete cycles (around $1.5\,h$), operation states with maximal temperature differences below 25 °C are reached after regeneration, whereas the deviation averaged along the reactor length represents only a difference of 5 °C. For the profiles at the end of the production periods, both maximal and averaged temperature deviations are below 5 °C.

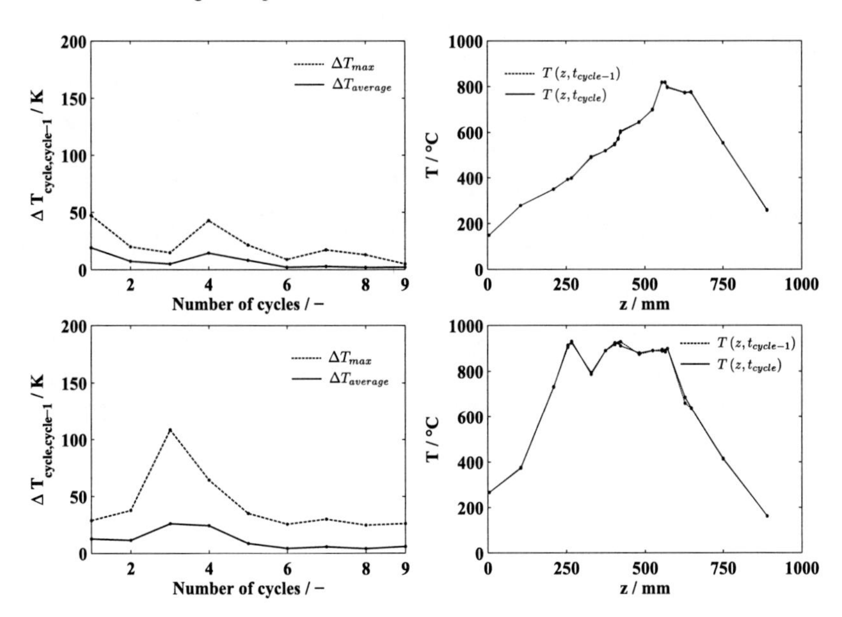

Figure 5.7: Transition to cyclic stationarity during experimental operation. Maximal and average temperature differences between consecutive end temperature profiles after production (top, left) and regeneration (bottom, left). Temperature profiles over the reactor length after production (top, right) and regeneration (bottom, right) in stationary state.
Operation conditions: $2.4\,kW_{LHV,H_2}$; $\dot{V}_{inert} = 15\,Nl\,N_2/min$; $\tau_{prod} = \tau_{reg} = 300\,s$.

Comparison of the end profiles after regeneration provides normally larger differences than after production since the exothermicity and severity of the reaction conditions during combustion have not the smoothing effect of the endothermic reaction on the temperature profiles. As it can be extracted from the graphs on the right side of figure 5.7, the operation state is, under such conditions, stable enough to enable a quantitative interpretation of the reactor behaviour. After such a state is reached, variations in the operation parameters induce perturbations that are generally dampen within $0.5\,h$.

5.3 Experimental results

The results presented in this section verify the applicability of the reheating strategy discussed in the previous chapters to efficiently regenerate the RFR under periodic operation. Furthermore, insight into its behaviour and the sensitivity towards relevant operation parameters, as well as into the reactor performance are gained.

A detailed description of RFR operation with the previous setup has been previously reported by H. Dieter and Glöckler in [62]. The results reported here are discussed in detail to clarify and emphasize the characteristics and attributes of the new system. Special attention is paid to the robustness of operation under load changes and to the possibility of running the RFR using different fuels.

5.3.1 Performance indicators

Relevant aspects describing the RFR performance are the hydrogen yield and the thermal efficiency. The former is directly proportional to methane conversion, which can be determined analyzing the composition of the reactor effluent. Kinetic measurements carried out in previous work prove the outstanding activity of the catalyst used and suggest that the system is not subject to kinetic limitations except for the water-gas shift reaction (eq. 2.2), limited at temperatures below 450 °C [102]. Thus, the product composition can be expected to coincide with that under equilibrium conditions at the temperature of the gas leaving the catalytic section of the packed bed.

The thermal efficiency, in turn, can be defined as the relation between the consumed energy during production and that supplied during regeneration:

$$\eta_{therm} = \frac{Q_{prod}}{Q_{reg}} \cdot 100 \approx \frac{\dot{N}^{+}_{CH_4} \cdot \overline{X}_{CH_4} \cdot \left|\Delta h_{ref}\right| \cdot \tau_{prod}}{\sum\limits_{CC=1}^{3} \sum\limits_{t=0}^{\tau_{reg}} \left(\dot{N}_{fuel,CC} \cdot \Delta t\right) \cdot \left|\Delta h_{comb}\right|} \cdot 100 \tag{5.1}$$

The term $|\Delta h_{ref}|$ accounts for the reaction enthalpies of both reforming (eq. 2.1) and water-gas shift (eq. 2.2) reactions and $|\Delta h_{comb}|$ for the combustion enthalpy of the fuel mixture under consideration. An accurate estimation of $|\Delta h_{ref}|$ requires information on the composition of at least two components present in the reactor effluent during production (CH_4 and CO or CO_2). The latter can be either experimentally measured or estimated based on the assumption of thermodynamic equilibrium for the gas leaving the catalytic section.

Within the experimental results next discussed, significant differences between efficiency evaluations based on experimental and on estimated data have been observed. Their origin cannot be unambiguosly defined, but is clearly affected by the limitations of the analytic system (cf. sec. 5.1). Moreover, possible deviations from the equilibrium assumption under certain operation conditions might further contribute to the divergence between the data evaluated. In spite thereof, the relative variations of the performance indicators induced by changes in the operation conditions are still a valid indicator to evaluate improvements or regressions in the performance of the RFR.

5.3.2 RFR behaviour characteristics

The characteristics of RFR operation already discussed during the introduction of the concept in section 2.3 are next analyzed based on experimental results. Stationary operation at a reforming load of $2.4\,kW_{LHV,H_2}$ and a cycle duration of 300 s is taken as reference. An inert volume flow $\dot{V}_{inert}$ of $15\,Nl\,N_2/min$ is used during regeneration, which is run with H_2 as fuel. The results discussed correspond to those represented in figure 5.7. Based on operation under this conditions, the reactor behaviour and the estimation of the reactor performance indicators is next detailed.

Production period: temperature and composition evolution

The temperature profiles along the reactor length under stationary operation are represented in figure 5.8 in intervals of 60 s. At the beginning of the production period, the temperature profile shows the main characteristics of an optimal initial profile. Low temperatures at both reactor ends contribute minimizing heat losses, whereas a constant, high temperature plateau over the length of the catalytic section between the leftmost and rightmost combustion chambers is established.

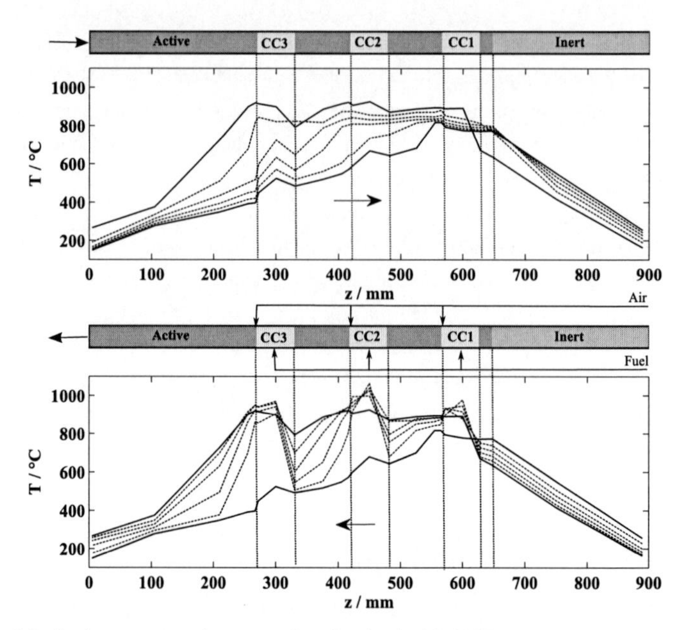

Figure 5.8: Stationary operation at a reforming load of $2.4\,\mathrm{kW_{LHV,H_2}}$. $\tau_{prod} = \tau_{reg} = 300\,\mathrm{s}$. Temperature profiles every 60 s for the production (top) and regeneration (bottom) periods.

The evolution of the temperature profile during the production period (fig. 5.8, top) depicts the behaviour described by Glöckler et al. in the theory [1]. Whereas the whole temperature profile is shifted in the flow direction as the period proceeds, the thermal front in the left reactor zone travels considerably faster than the front in the right (inert) reactor end. The latter behaves as a pure convective thermal wave. Thus, its velocity is directly proportional to the energy contained in the gas flow leaving the reactor, which in turn, depends on the mass flow, its composition and heat capacity. The velocity of the temperature profile in the temperature region above 500 °C is considerably higher than in lower temperature ranges. This effect can be identified in the evolution of the temperature slope in the rightmost inert section along the production period. Although the profile is solely measured with three thermocouples (beginning, middle and end of the inert section), maximal velocity differences of up to 7% can be estimated between the upper and lowest temperature regions. The temperature front becomes herewith somewhat steeper and thus, the efficiency of the bed as a regenerative heat exchanger can be slightly enhanced.

In the catalytic zone, the heat of reaction for reforming is supplied by the energy stored in the packed bed. Thus, energy consumption through reaction is coupled with convective energy transport, resulting in an increased velocity of the temperature front. Furthermore, as already discussed in section 2.3.1, the velocity of the reaction front strongly depends on the temperature in the packed bed section through which this front travels. Enhanced conversion in the temperature range above 700 °C translates in the formation of a steep profile similar to a shock wave. At lower temperature ranges, the thermal front behaves as a purely dispersive wave. These two different behaviours can be clearly identified in figure 5.9, which resembles the experimental results discussed in figure 2.4. The migration of a dispersive wave is measured over time at the reactor entrance (105 mm), which exhibits temperatures below 700 °C almost over the complete production period. On the contrary, between the leftmost and the middle combustion chambers, the migration of a steep reaction front characterized by a fast temperature decrease after approximately 2.5 min can be measured at position 375 mm.

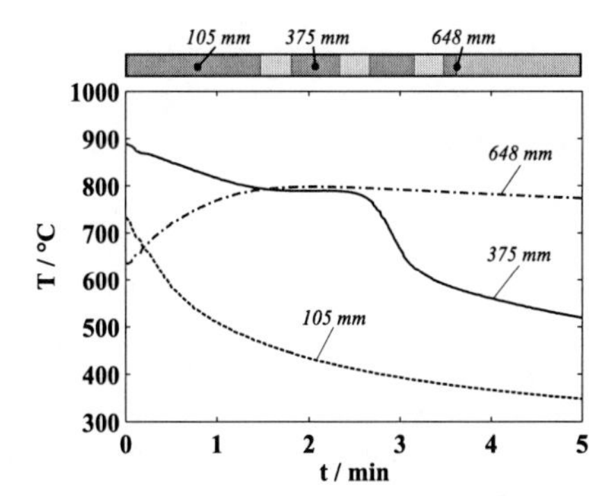

Figure 5.9: Temperature evolution at several positions along the reactor axis over a production period.

As already mentioned in section 5.3.1, the temperature at the outlet of the catalytic section during production can be taken as measure of the conversion achieved, given that thermodynamic equilibrium can be assumed. Accordingly, the temperature evolution at position 648 mm in figure 5.9 directly correlates with the reactor effluent composition, as shown in figure 5.10.

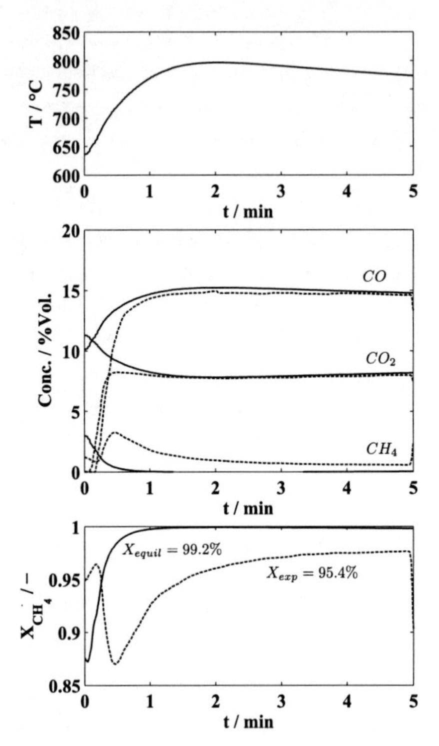

Figure 5.10: Top: temperature evolution over a production period at the oultet of the catalytic zone; middle: estimated (—) and measured (– –) CO, CO_2 and CH_4 composition at the reactor outlet; bottom: estimated (—) and measured (– –) methane conversion.

At the beginning of the production period, temperature at the outlet of the catalytic packing is far below that corresponding to large equilibrium conversion values. This is attributed to the cooling effect of the inert flow fed from the right end of the reactor during the preceeding regeneration period. As soon as the production period starts, temperature increases rapidly through convective heat transport. After approximately 1 minute, temperature reaches its maximum and, disregarding a slight temperature decrease attributed to heat losses of the system, this value remains almost constant during the rest of the period.

A comparison of the estimated equilibrium composition based on the temperature at the outlet of the catalytic zone and the composition measured with the NDIR shows a very good agreement of the data after the first production minute (figure 5.10, middle). The somewhat larger deviations at the beginning can be attributed to the dead times in the sampling configuration (cf. sec. 5.1) and to some extent, to the lack of a purge step between production and regeneration periods. These deviations are responsible for the difference between the equilibrium conversion X_{equil}, based upon the exit temperature of the last catalyst section (fig. 5.10, top), and the calculated conversion X_{exp}, based upon the exit concentration measurements (figure 5.10, bottom).

The results discussed above, question the adequacy of the reactor configuration and in particular, of the presence of the catalytic packing between the rightmost combustion chamber and the inert section of the reactor. The conversion achieved in the reactor is strictly limited by the temperature of the latter, which is kept at the temperature required to achieve high conversion by convective heat transport during the production period. However, it is cooled down during regeneration if the (inert) regenerative heat exchanger section turns to be underdimensioned. This effect can be clearly seen in figure 5.8, bottom, where the temperature in the rightmost catalytic zone significantly decreases during the regeneration period. Thus, it can be stated that the presence of this zone strongly limits the optimization of the operation parameters. This should be considered in future studies.

Regeneration period using H_2 as fuel: suppression of excess temperatures

The flow is reversed and regeneration starts as soon as the temperature at the right reactor outlet reaches values above 250 °C. Table A.6 summarizes, amongst other parameters, the real factors applied to the fuel/air flows in order to counteract heat losses. As expected, the real factor decreases in the direction of the flow, i.e. from the first combustion chamber (CC_1) to the last one (CC_3), due to the increase in the mass flow rate after each chamber (cf. sec. A.1.4).

Since H_2 exhibits low ignition temperatures, combustion takes place homogeneously in every combustion chamber from the very beginning of the regeneration period. Comparison of figures 2.9 in chapter 2 and figure 5.8, reveals some of the advantages of the current reheating strategy. For instance, minimization of the thermal mass in the combustion sections translates into a smoothing effect on the temperature profile. This can be seen in figure 5.8, bottom, where the temperature peaks measured in the chambers CC_1, CC_2 and CC_3 during regeneration disappear within a few seconds, as soon as fuel/air supply to the chambers is interrupted, prior to the subsequent production period.

Figure 5.11 provides a detailed insight into the behaviour of the combustion chambers during the regeneration period. The temperature evolution bears witness to the simplicity of the control strategy and depicts the behaviour of the reactor during regeneration. The control algorithm of each combustion chamber is designed to stabilize the chamber temperature by manipulating the air and fuel flow introduced. The rightmost combustion chamber CC_1 serves essentially, in the present reactor configuration, to balance the heat losses in the system. The cooling effect of the inert flow entering the reactor from the right end increases the energy demand as the regeneration period proceeds. It becomes again evident, that the inert regenerative heat exchanger section is underdimensioned for the current operation conditions, affecting the thermal efficiency of the concept.

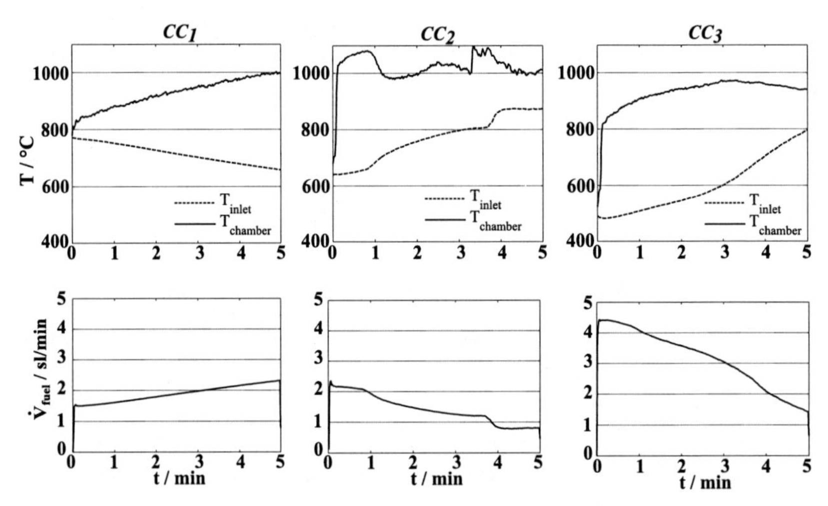

Figure 5.11: Top, from left to right: temperature evolution at the inlet and in chambers CC_1 to CC_3 respectively during regeneration. Bottom, from left to right: fuel volume flow fed to chambers CC_1 to CC_3 during regeneration subject to the control strategy.

The chamber in the middle of the packed bed shows the most dynamic behaviour. During the first minute of the regeneration period, the chamber operates at its maximal load. As soon as the thermal wave generated in the first combustion chamber reaches the second chamber and the inlet temperature increases, the amount of fuel/air supplied decreases in order to keep the temperature within the maximal value of $T_{max} = 1000\,°C$ as stable as possible. During the last regeneration minute, the packing section between the first and second chambers has reached its stationary temperature profile and the second chamber simply compensates the difference between the inlet temperature and the set point T_{max}. The strong temperature oscillations in the second chamber contrast with the smooth profiles of the fuel flow and highlight the sensitivity of the combustion conditions in the chamber towards small changes in the operation conditions and thus, in the flow patterns established. Although the maximal temperatures reached remain far below the limits set by the mechanical stability of the construction materials, the dynamics of the chamber temperature could be dampened by introduction of a time dependant *real factor* and an integral term in the control algorithm underlying the fuel/air supply.

The leftmost chamber is supplied with almost twice as much fuel as the first and second chambers. Two reasons contribute to this effect. The inlet temperatures exhibit the lowest values compared to chambers CC_1 and CC_2, since the endothermic reaction front consumes the energy stored in the packed bed from left to right during the production period. Furthermore, the mass flow fed to the reactor during the regeneration period is increased in the previous chambers through the supply of fuel and combustion air, translating in an increased energy input requirement to achieve a given temperature increase $\Delta T = T_{max} - T_{in}$. Accordingly, the largest increase in the heat flux in flow direction during regeneration occurs in the last combustion chamber.

The regeneration period reaches its end and the flow is reversed as soon as the temperature at the leftmost reactor end exhibits values above 250 °C. Larger temperatures are avoided during operation of the experimental setup, not only due to a decrease in the thermal efficiency of the system, but to avoid damaging of the reactor periphery. Based on the operation discussed in the current section, thermal efficiencies of $47.0 \pm 1.8\%$, as defined in section 5.3.1, have been reached. These are close to the average ones in conventional reforming technologies. However, considering the reduced dimensions of the experimental reactor, the large heat losses to which the system is subject and the fact that neither the operation nor the reactor configuration have been optimized up to this stage of the work, the degree of heat utilization obtained can be graded as very promising.

5.3.3 Operability and robustness of operation

One of the targets pursued with the current experimental setup is to prove the robustness of operation of the RFR concept. The latter is a fundamental feature that the reactor concept should exhibit, providing versatility towards different productivity requirements. Two relevant operation parameters, fuel used during regeneration and reforming load, have been varied within this context. The reactor response to these perturbations is analyzed next.

Fuel variation

The analysis discussed next considers PSA off-gas as fuel and takes operation with hydrogen as fuel and at the same reforming load of $2.4\,kW_{LHV,H_2}$ and $\tau = 300\,s$ as reference (cf. fig. 5.8). Fuel variations may have a large impact in the reactor operation for obvious reasons. The heating value and thus, the amount of fuel required depends strongly on the composition of the latter. Simultaneously, the combustion chemistry directly affects the amount of oxygen/air required for combustion. Accordingly, the heat fluxes along the packed bed are affected by the fuel of choice.

Based on the procedure detailed in section 5.2.2 for a given reforming load, it can be estimated that the use of PSA off-gas with the composition given in table 2.1 requires around 20% less inert flow as well as an increase of almost 8% of the fuel used compared to regeneration with H_2 (cf. tab. A.5 and A.2). In the current example, the inert flow has been reduced by almost 30% and the total fuel amount increased by 11% in comparison to the reference periodic stationary state, achieved with hydrogen for a reforming load of $2.4\,kW_{LHV,H_2}$.

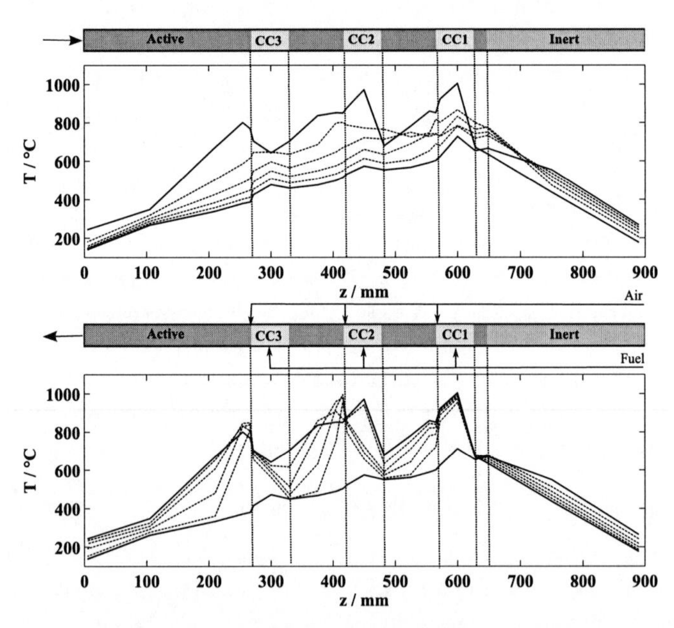

Figure 5.12: Temperature profile evolution during regeneration with PSA off-gas as fuel for a reforming load of $2.4\,kW_{LHV,H_2}$ and $\tau = 300\,s$.

Since synthetic PSA off-gas exhibits a significantly larger homogeneous ignition temperature than H_2, only if the inlet temperature of the gas entering a chamber exceeds a temperature minimum, homogeneous combustion is to be expected. Figure 5.12 depicts this behaviour. The bulk of the rightmost chamber ignites from the very beginning of the regeneration period, whereas the central chamber serves as mixing device during the first half of the period and catalytic combustion takes place downstream. As the gas inlet temperature reaches values close to 700 °C, the mixture ignites homogeneously. In the leftmost chamber, the above mentioned temperature is only reached at the end of the regeneration period, so that in essence, only catalytic combustion takes place.

The behaviour of the central combustion chamber delivers some further interesting information. As soon as homogeneous combustion occurs, the reheating temperature level in the packing decreases. Under heterogeneous combustion conditions, temperature measurements within the first packing centimeters are a proper indicator of the conditions in the reaction zone. On the contrary, as already discussed in chapter 4, as soon as homogeneous reaction prevails, the maximal temperature measured in the chamber might not be representative for the average temperature of the bulk, depending on the formation of segregated reaction zones. Provided that the chambers are subject to larger heat losses than the packing sections and that the fuel/air flow rates are controlled based on the maximal temperature measured, it can be expected that the average temperature of the gas leaving the chamber lays below the highest value measured. A similar behaviour can be also observed in the rightmost chamber and reveals the importance to enhance temperature homogeneity in the chamber, whenever homogeneous combustion occurs.

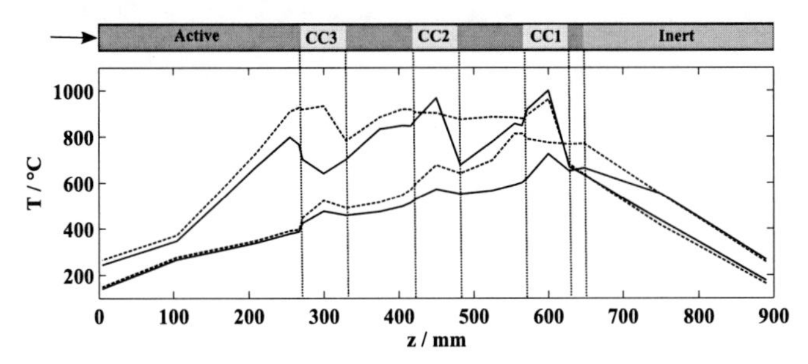

Figure 5.13: Temperature profiles at the beginning of the production and regeneration periods for steady state operation at a load of $2.4\,kW_{LHV,H_2}$ and $\tau = 300\,s$. Operation with H_2 (– –) and PSA off-gas (—) as fuel.

Besides the differences during the regeneration period, figure 5.13 enables a direct comparison of the temperature profiles under steady state for operation with H_2 and PSA off-gas respectively. In an attempt to maintain the maximal temperatures at the reactor exit boundaries at about 200 °C independently of the fuel used, the heat flux during regeneration with PSA off-gas must be reduced. Decreasing the energy input into the leftmost chamber CC_3 by reduction of the real factor RF applied to the fuel/air flow rates has been the approach followed. However, the direct consequence of this action is a reduction of the energy supplied to the system. The heat losses in the leftmost chamber cannot be compensated and the set point for the reheating temperature is not reached by the end of the regeneration step. The energy content per unit volume of packing is herewith reduced and the distance travelled by the endothermic front during production larger than in the reference case. This, in turn, results in a reduced average temperature level in the active sections of the packed bed.

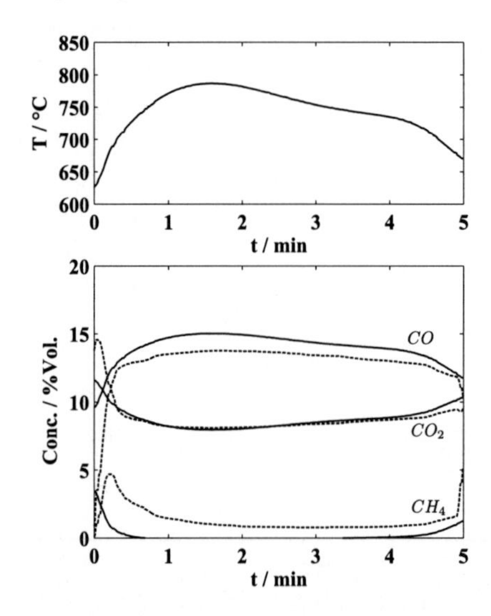

Figure 5.14: Temperature evolution over a production period at the oultet of the catalytic zone (top) and estimated (—) and measured (−−) CO, CO_2 and CH_4 composition at the reactor outlet (bottom) for steady state operation at a load of $2.4\,kW_{LHV,H_2}$, $\tau = 300\,s$ and reheating with PSA off-gas.

Although the chemical performance of the reactor is assumed to be independent from the fuel used, the lower average temperature causes a reduction of the average measured conversion from 95.7% in the case of regeneration with H_2, to 92.7% for operation with PSA off-gas. Furthermore, comparison of figures 5.10 and 5.14 reveals a significant deviation between the estimated and measured CO and CH_4 compositions for operation with PSA off-gas. Again, this effect is not to be attributed to the fuel used, but to a possible kinetic limitation of the reforming reaction due to the lower temperature level at the outlet of the rightmost catalytic zone.

Conversion decrease, together with slightly larger temperatures measured at the left reactor boundary during regeneration, negatively affect the thermal efficiency of the system, which falls from a maximum of 47.8% in the reference case to a value of 44.9% in the current one. Furthermore, the shape of the temperature profiles, with considerable temperature gradients in the active sections between the combustion chambers, highlights a suboptimal use of the catalytic packing and the energy stored in it. For a given set of operation parameters, i.e. reforming load and cycle duration, a uniform temperature plateau contributes enhancing the efficient use of the catalytic mass in the packing. In the current case, the temperature evolution during production depicted in figure 5.12, top, reveals that the lack of a uniform, high temperature level hinders the formation of a stable endothermic reaction front. The temperature profile evolves with a strong dispersive component, translating in a rapid temperature decrease in the rightmost catalytic zone and thus, in conversion and energetic efficiency losses.

Based on the comparison of regeneration with H_2 and PSA off-gas (cf. fig. 5.13), it can be stated that a uniform temperature plateau can be rather achieved with fuels exhibiting large heating values and a low ratio between the stoichiometric coefficients of oxygen and fuel. In such a case, the increase of the heat flux along the direction of the flow during regeneration and the heat losses through the reaction boundary can be minimized, improving the overall thermal efficiency of operation.

Reforming load variation

The reforming load can be raised or reduced changing the mass flow fed to the reactor during the production period. Accordingly, the energy supply during the regeneration period as well as the cycle duration need to be adjusted, so that operation remains stable and the temperature of the gas flow leaving the reactor does not exceed values that could damage the periphery of the setup.

At higher loads, energy consumption increases proportionally to the reaction extent. At the same time, due to the increased heat flux of the flow leaving the reactor, the period duration needs to be proportionally reduced in order to keep the heat losses through the reactor boundaries at a constant level. Thus, if the operation parameters set is varied accordingly to the load change, variations can be made without significantly disturbing the reactor performance.

To illustrate this, operation characterized by the temperature profiles presented in figure 5.8 is taken as reference, whereas the reforming load is increased by about 25% up to almost $3\,kW_{LHV,H_2}$. Simultaneously, in order to balance the operation parameters, the period duration has been reduced by 33% while the inert flow fed to the reactor during regeneration has been increased by 33%. After correction of the mentioned parameters (see comparison of the parameter sets in table A.6), operation remains almost invariable. Accordingly, after completion of one cycle, the system can be already considered to have attained periodic stationarity.

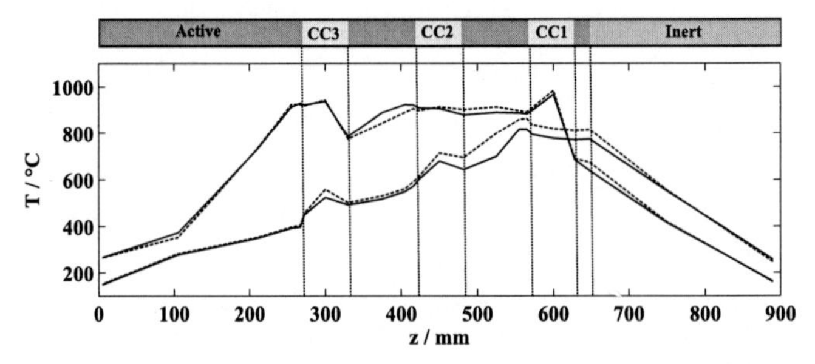

Figure 5.15: Temperature profiles at the beginning and the end of a production period for a load of $2.4\,kW_{LHV,H_2}$ and $\tau = 300\,s$ (—) and $3\,kW_{LHV,H_2}$ and $\tau = 200\,s$ (– –). Operation parameters summarized in table A.6.

Direct comparison of the temperature profiles at the beginning and at the end of the production period in figure 5.15 highlights the similarity of both operation configurations mentioned. The slight disproportion of the heat fluxes during production and regeneration causes minor differences between both temperature profiles under stationary operation. At higher reforming loads, the relative importance of heat losses to the surrounding is reduced, since these are load-independent. Furthermore, the increase in the heat flux during the production period results in a faster convective energy transport to the rightmost catalytic zone. This, together with the slightly higher energy supply during regeneration compared to that consumed during production, leads to an increased temperature level in the rightmost catalytic zones. Thus, with a measured value of 97.4%, a slightly higher average conversion than in the previous case is obtained. It is remarkable that in the leftmost catalytic zones, no temperature differences are noticeable at the end of the production period, independently of the production load. This suggests that chemical equilibrium is achieved in the inlet reactor zone, providing the temperature profile with a slight convex form due to heat release attributed to the WGS reaction.

Regarding the thermal efficiency of the setup, it can be stated that variations in the reforming load have not an enhancing or worsening effect if proportionality of the remaining operation parameters is maintained. For a reforming load of $3\,kW_{LHV,H_2}$, a thermal efficiency of 43.0 to 43.5% has been measured, which is in the same range than that obtained with the reference case (cf. section 5.3.2).

As for operation with H_2, load variations in the range discussed above using PSA off-gas do not significantly affect the reactor performance whenever the relevant operation parameters are adjusted in order to balance the differences induced. Results corroborating this observation are summarized in table A.7. However, significantly increasing the operation load, e.g. doubling it up to $5\,kW_{LHV,H_2}$, highlighted the operation limits of the current experimental setup, which are next reported.

Operation limits

Figure 5.16 depicts stationary temperature profiles for operation at a reforming load of $4\,kW_{LHV,H_2}$. The duration of the production period has been accordingly adjusted to 100 s. Provided that the combustion chambers in the current setup exhibit a better performance in a narrow range of volume flow rates (cf. section 4.2), the duration of the regeneration cycle has been kept at 200 s in order not to deviate from the convenient operation range. Details on the parameters set are summarized in table A.7.

As it can be extracted from the temperature evolution during production (fig. 5.16, top), larger conversions than in the previous cases discussed are to be expected, provided that the temperature at the rightmost catalytic zone is permanently above 800 °C. An average conversion of 99.6% is estimated based on the temperature information. This corresponds to an average thermal efficiency of 54.4%, which is significantly larger than that reported for the reference case studied (cf. section 5.3.2). However, the analysis of the product gases reveals a significantly lower conversion than that assumed under equilibrium conditions. Conversion determined experimentally is 92.5%, so that the measured thermal efficiency only accounts for 42.4%.

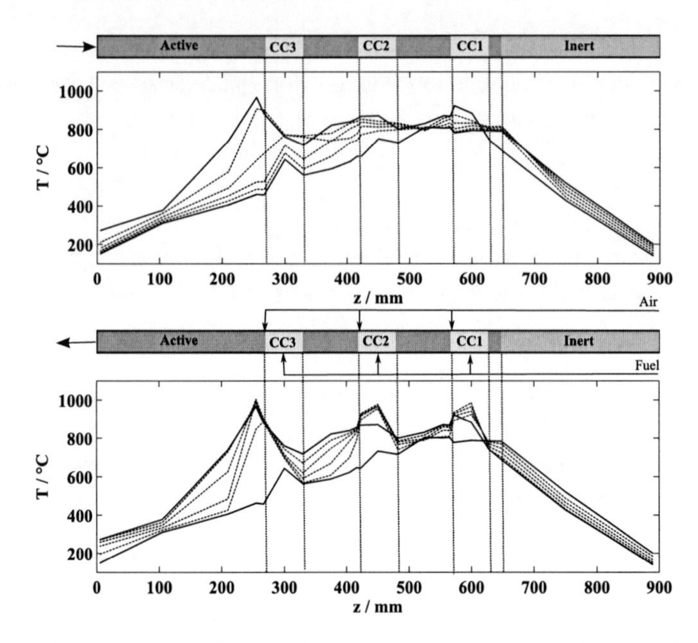

Figure 5.16: Temperature profile evolution during production (top) and regeneration (bottom) for a reforming load of $4\,\mathrm{kW}_{\mathrm{LHV,H_2}}$. $\tau_{prod.} = 100\,s$ and $\tau_{reg.} = 200\,s$. Use of PSA off-gas as fuel.

A decaying trend of the measured conversion has been observed for increasing productivities. Figure 5.17 illustrates this effect on a basis of the *GHSV* at the inlet and oulet conditions of the flow, for several reforming loads tested. While the average equilibrium and measured conversions differ around 7.1% for the reported load of $4\,\mathrm{kW}_{\mathrm{LHV,H_2}}$, this difference becomes significantly larger by increasing the reforming load to 5 and $5.5\,\mathrm{kW}_{\mathrm{LHV,H_2}}$. In such cases, the difference between estimated and measured conversions, defined as ΔX, reaches values of 9.1% and 12.4% respectively.

The cause of these strong deviations can be attributed to several aspects. In an attempt to determine the reforming kinetics on the current catalyst reported by Gose in [102], no kinetic limitations were observed even at significantly larger GHSV values than those reported in figure 5.17. However, on the basis of the current results, mass transfer limitations, possibly amplified by catalyst deactivation effects, cannot be ruled out. Furthermore, formation of preferential ways or flow-bypass within the reactor due to the strong mechanical stresses and material expansion to which the construction has been submitted during operation cannot be excluded either. Within the scope of this work, not enough evidences helping to clarify the origin of this behaviour have been gathered. Hence, this aspect should be addressed in detail in future studies.

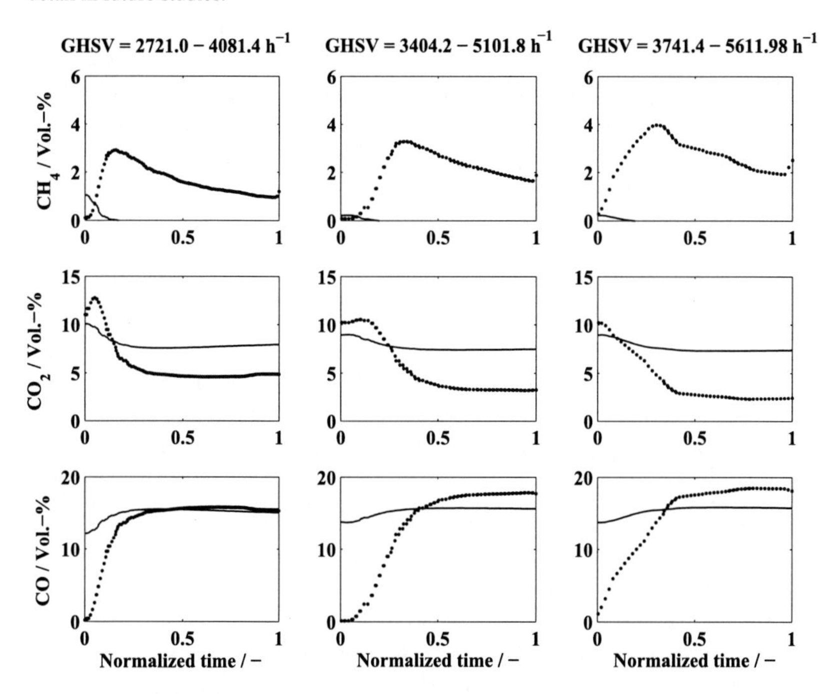

Figure 5.17: Measured (•) and equilibrium concentrations (—) of methane (top), carbon dioxide (middle) and carbon monoxide (bottom), during reforming at 4 (left), 5 (middle) and 5.5 kW_{LHV,H_2} (right) respectively.

Even if the limitations arising from the productivity loss at higher loads is obviated, figure 5.16 further reveals a suboptimal use of the catalyst between the leftmost and rightmost combustion chambers, CC_3 and CC_1. Particularly in the zone between chambers CC_2 and CC_1, energy released during regeneration merely counteracts the heat losses of the system and the cooling effect of the carrier gas fed to the reactor inlet. Comparing the maximal temperatures reached at both reactor boundaries, it becomes clear that heat flux disproportion during production and regeneration contributes to the poor use of the catalyst. During regeneration, the left reactor boundary reaches temperatures close to 300 °C, whereas the right boundary reaches maximal values of 200 °C during production. Thus, extending the duration of the production period or reducing that of the regeneration one represents an approach to improve the catalyst use and to help balance the heat fluxes between both periods. However, the new attainable stationary state is hardly predictable based only on experimental observations. This motivates the introduction of simulation tools in order to quantitatively estimate the potential of the actual reactor concept and to explore alternatives that might help improve the performance of the system. These aspects are discussed in detail in the following chapters.

5.4 Summary

Focus of this chapter is the validation, under periodic operation, of the reheating strategy developed within the scope of this work and reported in the previous chapters. Hence, the experimental setup and the RFR designed for this purpose are described. The results obtained verify the potential of the concept and prove that an efficient reheating at a high temperature level is possible, overcoming the limitations identified in the past, i.e. the occurrence of excess temperatures leading to structural damages in the packing and the reactor internals.

Variations of significant operation parameters, such as reforming load or the fuel used for regeneration, have been performed in order to corroborate the robustness of operation. Furthermore, their effect on the operation performance has been assessed based on indicators previously defined. Disproportion of heat fluxes during production and regeneration has been identified as a primary cause of operation efficiency losses. The magnitude of this disproportion and thus, the deviation from the best achievable performance for the reactor configuration considered, is sensitive to several parameters, e.g. the state dependency of the gas physical properties or the fuel used for regeneration. Simultaneously, the heat flux imbalance between periods can be partly counteracted through proper adjustment of some operation settings, and in particular of the regeneration-to-production mass flow ratio.

The complex interaction of the different operation parameters together with the limitations posed by the reactor design used, difficult an elucidating analysis of the effects caused by variations in the operation settings and the identification of an optimal performance range for the given reactor configuration on an experimental basis. Thus, the latter analysis as well as the proposal of further reactor configurations exploiting the potential of reverse-flow operation are pursued in terms of simulation studies and reported in the following chapters.

Chapter 6

Modelling the RFR

The current chapter describes the mathematical model used to analyze, on a simulation basis, the features characterizing the behaviour of the RFR reported in the previous chapter. Its main elements, i.e. the packed bed sections and the combustion chambers, are modelled independently and coupled for simultaneous simulation. As reported in chapter 7, this approach not only enables to describe the reactor design already discussed, but also exploring alternative reactor configurations and their performance.

6.1 Mathematical model

The current section addresses the assumptions underlying the mathematical model of the RFR and summarizes the mass and energy balances used to describe the system. Details on the model parameters such as physical and transport properties as well as criteria defining the simulation parameters set are detailed in appendix C.

6.1.1 Packed bed

Based on the reactor design described in section 5.1, almost 80% of its length is constituted by packed bed sections. The mathematical model used to describe them consists on a dynamic, heterogeneous, one-dimensional approach, accounting for the main processes taking place, i.e. heat generation and consumption as well as composition changes through chemical reaction and energy transport, accumulation and exchange within different phases. The system balanced is sketched in figure 6.1 and considers the solid packing and the gas bulk, as well as the reactor wall and an insulation layer, in order to account for their non-negligible thermal capacity and conductivity effects.

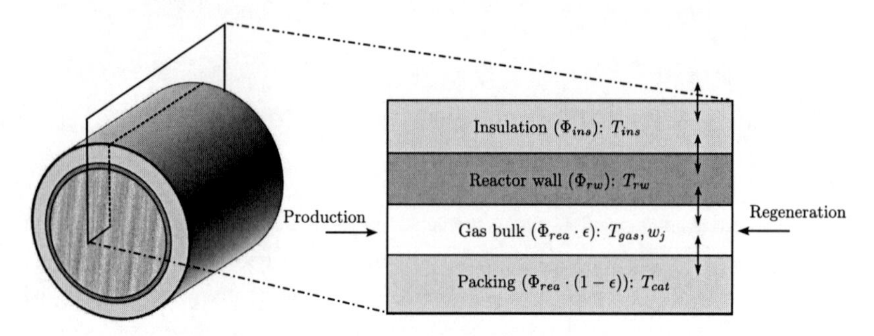

Figure 6.1: Schematic representation of a control element of a packed bed section. Each phase and the respective state variables balanced are listed. Phases are coupled through (heat) exchange with each other.

Model assumptions

For the description of the packed bed sections, following simplifying assumptions have been made:

- Ideal plug-flow behaviour is considered. Thus, radial temperature and concentration profiles are neglected.
- An isobaric system is considered. Pressure drop along the packed bed is neglected.
- The gas phase behaves ideally and its physical properties are given as functions of temperature and composition.
- Mass and energy balances in the gas phase account for convective and dispersive transport. In the solid phase, only conductive transport is considered. Heat and mass transfer within phases is formulated as sinks or sources, described in terms of linear driving forces (LDF).
- Heat transfer between the reactor wall and the packing is neglected. Only convective heat transport between the gas phase and the reactor wall is considered.
- All balances are defined referring to the total reactor volume. This is estimated accounting for the outmost diameter. Four different phase fractions can be defined: Φ_{ins} for the insulation, Φ_{rw} for the reactor wall and Φ_{rea} for the reaction zone, which in turn, comprises packing ($\Phi_{cat} = \Phi_{rea} \cdot (1-\varepsilon)$) and gas ($\Phi_{gas} = \Phi_{rea} \cdot \varepsilon$) phase.
- External mass transfer resistances are neglected.
- The physical parameters of the solid phases are considered to be constant.

Model equations of the bulk phases

Basis of the mass and energy balances used is the derivation of the conservation laws for one-dimensional distributed systems given in [103]. The different balances used to model the packed bed are next briefly described.

Solid phase energy balance $\left[kW/m^3\right]$:

$$\Phi_k \rho_k c_{p,k} \cdot \frac{\partial T_k}{\partial t} = \frac{\partial}{\partial z}\left(\Phi_k \cdot \lambda_k \cdot \frac{\partial T_k}{\partial z}\right) + q_{exchange} \tag{6.1}$$

The formulation in equation 6.1 applies for the insulation and the reactor wall, both referred by the subscript k. The term $q_{exchange}$ represents the volume based heat flux exchanged between phases. The reactor wall heat capacity is always accounted for, whereas that of the insulation can be optionally neglected. Based on this considerations the exchange term for the insulation phase is formulated as

$$\begin{aligned} q_{exchange} = {} & a^v_{rw,ins} \cdot k^{rw,ins} \cdot (T_{rw} - T_{ins}) \\ & - a^v_{ins,amb} \cdot k^{ins,amb} \cdot (T_{ins} - T_{amb}), \end{aligned} \tag{6.2}$$

for the reactor wall with insulation layer as

$$\begin{aligned} q_{exchange} = {} & a^v_{gas,rw} \cdot k^{gas,rw} \cdot (T_{gas} - T_{rw}) \\ & - a^v_{rw,ins} \cdot k^{rw,ins} \cdot (T_{rw} - T_{ins}), \end{aligned} \tag{6.3}$$

or without insulation layer as

$$\begin{aligned} q_{exchange} = {} & a^v_{gas,rw} \cdot k^{gas,rw} \cdot (T_{gas} - T_{rw}) \\ & - a^v_{rw,amb} \cdot k^{rw,amb} \cdot (T_{rw} - T_{amb}) \end{aligned} \tag{6.4}$$

The terms $a^v_{n,m}$ and $k^{n,m}$ refer respectively to the specific heat transfer area in m^2/m^3 and the heat transfer coefficient in $kW/m^2/K$ between phases n and m (see app. C.1.1). Under adiabatic conditions, the last term in equations (6.2) and (6.4) is set to zero.

Energy balance of the packed bed $\left[kW/m^3\right]$: The energy balance of the packing can be considered as an extension of equation (6.1) to additionally account for the energy consumed or released through chemical reaction.

$$
\begin{aligned}
\Phi_{rea} \cdot (1-\varepsilon) \cdot \rho_{cat} c_{p,cat} \cdot \frac{\partial T_{cat}}{\partial t} = & \frac{\partial}{\partial z}\left(\Phi_{rea} \cdot (1-\varepsilon) \cdot \lambda_{cat} \cdot \frac{\partial T_{cat}}{\partial z}\right) \\
& + \Phi_{rea} \cdot \alpha_{cat} \cdot a^{v}_{cat} \cdot (T_{gas} - T_{cat}) \\
& + \Phi_{rea} \cdot (1-\varepsilon) \cdot \sum_{i=1}^{3} \left[(-\Delta h_{r,i}) \cdot r_i^{cat}\right]
\end{aligned} \tag{6.5}
$$

The last term has to be considered only for catalytic active packing during the production period. In all other cases, the term is set to zero. Accordingly, the subindex i refers to the reforming, water-gas shift and direct reforming reactions respectively.

Gas phase energy balance $\left[kW/m^3\right]$:

$$
\begin{aligned}
\Phi_{rea} \cdot \varepsilon \cdot \rho_{gas} c_{p,gas} \cdot \frac{\partial T_{gas}}{\partial t} = & \frac{\partial}{\partial z} \Phi_{rea} \cdot (\dot{m} c_{p,gas}) \cdot \frac{\partial T_{gas}}{\partial z} \\
& \frac{\partial}{\partial z}\left(\Phi_{rea} \cdot \varepsilon \cdot \lambda_{eff} \cdot \frac{\partial T_{gas}}{\partial z}\right) \\
& - \Phi_{rea} \cdot \alpha_{cat} \cdot a^{v}_{cat} \cdot (T_{gas} - T_{cat}) \\
& - a^{v}_{gas,rw} \cdot k^{gas,rw} \cdot (T_{gas} - T_{rw})
\end{aligned} \tag{6.6}
$$

Gas component mass balance $\left[kg/m^3\right]$: The reaction system considered is mole increasing. Thus, an appropriate way of formulating the balances is on a mass fractions basis.

$$
\begin{aligned}
\Phi_{rea} \cdot \varepsilon \cdot \rho_{gas} \cdot \frac{\partial w_j}{\partial t} = & - \Phi_{rea} \cdot \dot{m} \cdot \frac{\partial w_j}{\partial z} \\
& + \Phi_{rea} \cdot \varepsilon \cdot \frac{\partial}{\partial z} \cdot \left(D_{eff} \cdot \frac{\partial w_j}{\partial z}\right) \\
& + \Phi_{rea} \cdot (1-\varepsilon) \cdot MW_j \cdot \sum_{i=1}^{3} \nu_{i,j} \cdot r_i^{cat}
\end{aligned} \tag{6.7}
$$

The term j designates the five compounds under consideration: CH_4, H_2O, H_2, CO and CO_2. The effective dispersive coefficient D_{eff} is assumed to be identical for all components. Mass balances for the packed bed have been obviated, provided that the storage capacity of the packing can be neglected.

Boundary conditions

Each of the one-dimensional partial equations described requires two boundary conditions for its solution. The latter derive from balancing mass and energy in a volume element at the inlet and outlet boundaries. The solid phases (*ins*, *rw* and *cat*) are considered to be ideally isolated beyond the integration domain. Thus, the boundary conditions are written as follows:

$$\left.\frac{\partial T_k}{\partial z}\right|_{inlet} = \left.\frac{\partial T_k}{\partial z}\right|_{outlet} = 0 \tag{6.8}$$

On the contrary, in the case of the gas phase, accounting for the gas flow and for dispersive effects, Danckwerts type boundary conditions can be written. For the energy these can be written as

$$\Phi_{rea} \cdot \dot{m} \cdot c_{p,gas} \cdot T_{gas}\big|_{inlet} - \Phi_{rea} \cdot \varepsilon \cdot \lambda_{eff} \cdot \left.\frac{\partial T_{gas}}{\partial z}\right|_{inlet} = \Phi_{rea} \cdot \dot{m} \cdot c_{p,gas} \cdot T_{gas}^{+} \tag{6.9}$$

$$\left.\frac{\partial T_{gas}}{\partial z}\right|_{outlet} = 0 \tag{6.10}$$

and for the mass balances

$$\Phi_{rea} \cdot \dot{m} \cdot w_j\big|_{inlet} - \Phi_{rea} \cdot \varepsilon \cdot D_{eff} \cdot \left.\frac{\partial w_j}{\partial z}\right|_{inlet} = \Phi_{rea} \cdot \dot{m} \cdot w_j^{+} \tag{6.11}$$

$$\left.\frac{\partial w_j}{\partial z}\right|_{outlet} = 0 \tag{6.12}$$

Reaction kinetics

The present model accounts only for the kinetics of the reforming reactions (eqs. (2.1) to (2.3)) occurring during the production period. The approach used is the one proposed by Xu & Froment [104], which is extensively used and reported in the literature [48–50, 105–107]. Furthermore, previous studies conducted at the ICVT have shown a satisfactory agreement between experimental and simulation results [58]. The estimation of the temperature dependent reaction enthalpies is based on data from [108].

During the regeneration period, combustion of methane, carbon monoxide and hydrogen (eqs. (2.7) to (2.9)) are considered. These reactions can take place both heterogeneously on the catalyst surface and/or homogeneousy in the gas phase. In the current model, catalytic combustion is deliberately ommited. Hence, during the regeneration period the packed bed acts as an inert energy storage system. The energy is released through combustion in the combustion chambers and convectively transported to the packing (cf. s. 6.1.2).

According to this assumption, the number of component mass balances in the gas phase can be significantly reduced. Based on equations (2.1) to (2.3), only two key components need to be be balanced, whereas the rest can be described through linear combination of the latter. A third component is further balanced in order to be able to fully describe the combustion reactions during the regeneration period in case that the current model would be extended to account for the catalytic combustion of the regeneration fuel gas. Thus, CH_4, H_2O and CO_2 are balanced following equation (6.7), while the remaining chemical species are described as follows:

$$w_{H2} = \frac{MW_{H2} \cdot \left(\dot{N}^{+}_{H_2} + \Delta\dot{N}_{CO_2} - 3\Delta\dot{N}_{CH_4}\right)}{\dot{m}} \tag{6.13}$$

$$w_{CO} = \frac{MW_{CO} \cdot \left(\dot{N}^{+}_{CO} - \Delta\dot{N}_{CH_4} - \Delta\dot{N}_{CO_2}\right)}{\dot{m}} \tag{6.14}$$

$$w_{N2} = 1 - \sum_{j=1}^{5} w_j \tag{6.15}$$

The chemical species indexed with j are thereby CH_4, CO, CO_2, H_2O and H_2.

6.1.2 Combustion chamber

The combustion chamber represents an empty section within two packed bed sections. Assuming full adiabaticity and perfect mixing, it can be modelled as a continuously stirred reactor. Figure 6.2 represents the system balanced and the variables considered in general terms. During the production period, the gas stream flows through the empty section without experiencing changes neither in the energy, nor in the mass content of the flow, i.e. $\dot{M}_{in}$ and $T_{gas,in}$ equal $\dot{M}_{out}$ and $T_{gas,out}$. During regeneration the direction of the flow is reversed and besides the main flow, air ($\dot{M}_{air}$) and fuel ($\dot{M}_{fuel}$) are fed to each of the chambers, where combustion reactions (2.7) to (2.9) occur. The mathematical model itself is next described.

Model assumptions

The most relevant simplifying assumptions made in the chamber modelling are next listed:

- The combustion chamber is adiabatic.
- The mixing of the incoming gases occurs spontaneously and is assumed to be perfect. The continuity balance is considered to be quasi-stationary.

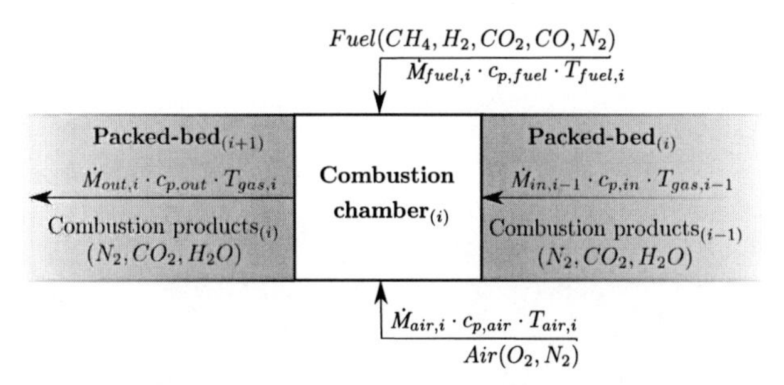

Figure 6.2: Schematic representation of a combustion chamber located between two packed bed sections and variables used to balance the system.

- A quasi-homogeneous, dynamic energy balance is considered. Accumulation in the gas (ε_{gas}) and in the chamber internals (ε_s) is accounted for.
- The reactants (air and fuel) are stoichiometrically fed into the chamber. Their supply undergoes a control strategy limiting the maximal temperature achieved by combustion (cf. sections 4.2.1 and C.1.5).
- Chemical reaction occurs spontaneously in the gas phase and is not kinetically limited.

Obviation of the kinetics governing the combustion reactions represents a noteworthy simplification of the model. The kinetic models generally used consist of complex elementary mechanisms, accounting for hundreds of radical generation and consumption steps even for the simplest hydrocarbons [109]. However, despite considerable efforts being made during the last decades to propose reduced mechanisms, the need for experimental validation and the additional complexity introduced is not justified at the current modelling stage.

Model equations

The combustion chamber is the only element in the reactor model which allows adding or draining a side stream to or from the main flow. The continuity equation is formulated as follows:

$$0 = \sum_{m=1}^{M} \dot{M}_{in} - \sum_{n=1}^{N} \dot{M}_{out} \tag{6.16}$$

As previously described and according to figure 6.2, equation (6.16) can be rewritten as $\dot{M}_{in} = \dot{M}_{out}$ during the production period. On the contrary, during the regeneration period, the flow entering the chamber is increased according to the supply of air ($\dot{M}_{air}$) and fuel ($\dot{M}_{fuel}$) for reheating purposes.

The gas phase energy balance in the chamber is described by equation (6.17):

$$\begin{aligned}(\varepsilon_{gas}\,\rho_{gas}\,c_{p,gas}+\varepsilon_s\,\rho_s\,c_{p,s})\,V_{cc}\,\frac{\partial T_{gas}}{\partial t} &= \dot{M}_{in}\cdot c_{p,in}\cdot T_{gas,in}\\ &-\dot{M}_{out}\cdot c_{p,out}\cdot T_{gas,out}\\ &+q_{combustion}\end{aligned} \tag{6.17}$$

The first term in the right side of equation (6.17) accounts for the energy contained in all incoming flows (cf. figure 6.2). During regeneration, the temperature of the air and fuel flows is set to a constant value of $T_{air,i} = T_{fuel,i} = 200\,°\mathrm{C}$. The temperature of the gas entering the chamber with the main flow $T_{gas,i-1}$ is set to that of the gas phase at the outlet boundary of the packed bed section upstream. The flow direction is accordingly taken into account.

The term $q_{combustion}$ in equation (6.17) represents the energy released by combustion. Thus, during the production period, the term is set to zero, whereas during the regeneration period it is estimated as follows:

$$q_{combustion} = \sum_{j=1}^{3}\left(-\Delta h_{C,j}\right)\cdot\frac{\dot{M}_{fuel,i}\cdot w_{fuel,j}}{MW_j}\cdot\underbrace{X_j}_{=1} \tag{6.18}$$

Thereby represents $\left(-\Delta h_{C,j}\right)$ the combustion enthalpy of the flammable components j = CH_4, CO and H_2 contained in the fuel mixture. The term X_j represents the molar conversion of each component, which based on the model assumptions can be obviated and set to one.

6.2 Numerical solution

DIANA is the simulation environment of choice to perform the study described in chapter 7. The name stands for Dynamic sImulation And Nonlinear Analysis and is developed as open source at the Max Planck Institute for Dynamics of Complex Technical Systems [110, 111]. One of is most interesting attributes is the possibility to build and run mathematical models in a modular way. The packed bed sections and the combustion chambers described in section 6.1 can be implemented as stand alone modules and coupled in multiple ways. It is thus possible to test different reactor configurations by changing their sequence or by making use of the numbering-up principle of any of the programmed modules that are part of a given modeling library. The modules itself are programmed in the model definition language *Mdl*, whereas models based on the available modules can be created with the open source modeling tool ProMoT [112, 113]. Although ProMot/DIANA are under continuous development and detailed documentation is only partially available, it has been widely tested so far and several examples describing its use and the details of the model implementation can be found [110, 114–116].

6.2.1 Model implementation

DIANA is conceived, amongst others, to perform stationary and dynamic simulations of chemical engineering processes that can be described as a system of nonlinear algebraic equations (NLE), or as initial-value problems in ordinary differential (ODE) or differential-algebraic (DAE) equations. SUNDIALS is the suite of numerical codes chosen for this purpose [117, 118]. The nummerical code IDA, based on Backward Differentiation Formula (BDF) methods, is the one used to solve initial-value problems for index-one DAE with the generic form

$$F(t, y, \dot{y}) = 0, \quad y(t_0) = y_0, \quad \dot{y}(t_0) = \dot{y}_0 \tag{6.19}$$

The partial differential equations (PDE's) describing the packed bed sections need to be discretized before being solved with DIANA. The spatial derivatives are approximated with finite differences along the integration domain, that consists of a mesh with an equidistant gridpoint distribution. Due to the steep temperature and concentration gradients ocurring during the RFR operation, an upwind second-order scheme for the first derivatives and a second-order centered scheme for the second derivatives has been used [119]. The boundary conditions represented by equations (6.8) to (6.11) can be implemented either as algebraic equations or, as in [120], introduced in the discretized form of the PDE resulting in an overall ODE system.

The models generated in ProMoT or directly in *Mdl* code are translated to C++ and further compiled before serving as input for the simulation environment DIANA. The latter is based on the programming language Python. The simulation scenarios can be thus defined and controlled with Python scripts [110]. Great advantage thereby is the possibility of making use of an extensive range of Python libraries, additionally to the own capabilities of the Python language as such.

6.2.2 Simulation control

Python scripts assume the simulation control. They are in charge of loading the mathematical model and building the data transfer interface between Python and the model itself, so that during simulation, a bidirectional data flow between user and model can take place. Furthermore, the scripts allow the choice of appropriate numerical solvers as well as their parameter adjustment. Graphic representation and data saving are also examples of the capabilities that Python adds to the simulation environment DIANA.

Periodic operation and stationary state

A peculiarity of the RFR mathematical model is the need to account for periodic operation and thus, for alternating boundary conditions and model parameters. Contrary to DIANA's predecessor DIVA, which enables the use of Petri nets to model discrete systems characterized by specific events, e.g. flow reversal, the former solely dealed with continuous models at the time of conducting the modelling task. Hence, the compiled model accounts for both operation modes, production and regeneration, which can be activated or deactivated via adjustment of appropriate flags. This is assumed by a Pyhthon script in charge of controlling the numerical solution of the model.

For given production and regeneration integration time spans, the script assumes the task of alternating between both operation regimes, initialising their integration and transferring the information to the subsequent periods. The state variables at the end of the regeneration period $y\left(t_{reg,end}, z\right)$ are adopted as initial value for the production period $y\left(t_{prod,0}, z\right)$ and vice versa. This strictly applies only to the state variables described by the energy balances (see previous section 6.1), fulfilling

$$T_k\left(t_{reg,end}, z\right) = T_k\left(t_{prod,0}, z\right) \tag{6.20}$$

whereas the concentration profiles of each chemical specie is updated after each regeneration step and prior to the initialisation of the subsequent production step, assuming thermodynamic equilibrium conditions for the given temperature and pressure profiles and a mixture of CH_4 and H_2O with $S/C = 3$

$$\begin{aligned} w_j\left(t_{reg,end}, z\right) &\neq w_j\left(t_{prod,0}, z\right) \\ w_j\left(t_{prod,0}, z\right) &= w_j^{eq.}(T_{gas}(t_{prod,0}, z), p(t_{prod,0}, z)) \end{aligned} \tag{6.21}$$

For the estimation of the equilibrium composition, the software toolkit CANTERA is used. The latter was conceived by David Goodwin, from the California Institute of Technology, to solve problems involving chemical kinetics, thermodynamics and transport processes [121] and can be integrated in the Python environment. This has proved to significantly enhance the computation performance and smoothened the reinitialisation step after integration of each regeneration step in order to satisfy the DAE residual

$$F(t_0, y_0, \dot{y}_0) = 0 \tag{6.22}$$

Due to the strong nonlinearity of the reaction kinetic terms in equation (6.5) and the strong interdependence between the several state variables, omitting this step is the cause of frequent initialisation problems.

Stationarity of periodic operation is assumed to be reached as the following condition is satisfied

$$max\left(y\left(t_{reg,end}\right)_i - y\left(t_{reg,end}\right)_{i-1}\right) \leq tol \tag{6.23}$$

Subscript i represents a counter for each complete cycle (production and regeneration), y the vector containing the value of the state variables at every position over the reactor length and tol the tolerance setting the maximal residuum accepted by the user ($1e-3$).

6.3 Performance indicators

The performance of different operation conditions and reactor configurations is analyzed based on indicators that give an account of the chemical and thermal efficiency of the system. **Methane conversion** during the production period is the parameter best defining the reactor productivity. Thus, average as well as maximal and minimal conversions during reforming are taken as indicators. The former is defined as follows

$$\bar{X}_{CH_4} = \frac{1}{\tau_{prod}} \int_0^{\tau_{prod}} X_{CH_4}(t)dt \tag{6.24}$$

Conversion achieved is determined by the temperature at the outlet of the catalytic section during the production period $T_{cat,out}$. Optimal operation conditions ensure a high, preferably constant conversion over the whole production period. Since at 1 bar pressure and around

850 °C full conversion is expected, a further interesting parameter is based on the so called equilibrium temperature T_{equil}. The latter represents the temperature required to match the conversion achieved, under the assumption that thermodynamic equilibrium is attained. This temperature can be determined for a given conversion, e.g. solving an expression of the form of equation (6.25), as used by Glöckler et al. in [1].

$$X_{CH_4} = X\left(p, T_{equil}\right) = \frac{e^{b(T_{equil}-a)+c(T_{equil}-a)^2+d(T_{equil}-a)^3}}{1+e^{b(T_{equil}-a)+c(T_{equil}-a)^2+d(T_{equil}-a)^3}} \tag{6.25}$$

Heat losses itself are considered in the definition of a parameter, which accounts for the efficiency of the heat integration in the RFR or its autothermicity. The **thermal efficiency** η_{therm} is an interesting indicator of the proportion of energy introduced into the system during the regeneration period that is consumed by the endothermic reaction. In an ideal autothermal system with perfect recovery of the heat capacity of the outcoming gases to preheat the incoming gases, themal efficiencies of $\eta_{therm} \approx 1$ should be expected. Based on such assumptions, the rigorous definition of η_{therm} is as follows:

$$\eta_{therm} = \frac{\int\limits_{0}^{\tau_{prod}} \int\limits_{0}^{z} (1-\varepsilon)\cdot A \cdot \sum\limits_{i=1}^{3} \left|r_i \cdot (-\Delta h_{R,i})\right| dz\, dt}{\int\limits_{\tau_{prod}}^{\tau_{reg}} \frac{m_{fuel}(t)}{MW_{fuel}} \cdot \left|(-\Delta h_{C,fuel})\right| dt} \tag{6.26}$$

The definition of η_{therm} according to equation (6.26) does not consider the energy required to heat up the regeneration gases (fuel, air and inert gas) entering the combustion chamber. As mentioned in section 6.1.2, fuel and air are supplied to the system at a temperature of 200 °C in the simulation studies. Accordingly, even under adiabatic conditions, thermal efficiencies below 100% are to be expected.

Besides this effect, the energy contained in the effluents that is not recovered in the regenerative heat exchanger sections at both reactor ends, is the main detriment of η_{therm} in an adiabatic system. **Average outlet temperatures** during production ($\bar{T}_{ref}$) and regeneration ($\bar{T}_{reg}$) are indicators of energy contained in the effluents. Since the heat capacity of the outgoing gas flow during production is almost three times larger than during the regeneration period, an averaging parameter as the one defined by Kolios in [122] to describe the **adiabatic temperature increase** of the gases leaving the reactor during a complete operation cycle (production and regeneration) is used in order to complement the interpretation of the results. The average adiabatic temperature increase of the product flow is defined as

$$\overline{\Delta T_{ad}} = \frac{1}{\tau_{prod}} \int_{0}^{\tau_{prod}} (T_{outlet} - T_{inlet})|_{prod} \cdot dt + \frac{1}{\tau_{reg}} \frac{\overline{(\dot{M} \cdot c_{p,reg})}\Big|_{\tau_{prod}}^{\tau_{reg}}}{\overline{(\dot{M} \cdot c_{p,prod})}\Big|_{0}^{\tau_{prod}}} \int_{\tau_{prod}}^{\tau_{reg}} (T_{outlet} - T_{inlet})|_{reg} \cdot dt \quad (6.27)$$

Chapter 7

Simulation study

The reactor configuration used in the experimental setup is based on the conceptual design described by Glöckler et al. in [1, 39], conceived to be operated in an asymmetric, periodic, production/regeneration mode. This design relies on the assumption of zero thermal creep velocity difference between production and regeneration cycles, which establishes only under equalty of heat flux capacities in both directions [37, 60]. However, deviation from this assumption induced by the state dependency of the flow properties as well as by the reheating concept newly developed, may affect an optimal energetic coupling of the endothermic and exothermic periods.

Hence, the first section of this chapter identifies those parameters significantly intensifying the mentioned heat flux disproportion. In turn, those operational adjustments that minimize or counteract the drawbacks of the system are briefly discussed. Furthermore, the modelling tool used and already introduced in chapter 6, is applied in order to consider alternative operation modes and reactor configurations that have not been yet experimentally tested and might represent an alternative to the asymmetric operation pattern considered so far.

7.1 Asymmetric operation

7.1.1 Convective heat flux disproportion

Convective heat flux is a function of the mass flow and the state dependency of the heat capacity of the gas phase, which exhibits a strong dependency on temperature and composition under relevant conditions for the production and regeneration periods. This dependency is summarized in figure 7.1. The increase of heat capacity with temperature causes an acceleration of the thermal wave at higher temperatures, enhancing the dispersive behaviour of the thermal fronts. This effect is particularly pronounced during the production period, due to composition changes experienced by the reacting flow. The heat capacity of the effluent is almost 30% larger than that of the feed (cf. fig. 7.1, dotted and straight lines in the left representation), so that energy losses beyond those originally predicted must be taken into account even though the mass flow density remains unchanged. To a lower extent, the same behaviour occurs during the regeneration period, given that the inlet flow under experimental conditions consists of pure nitrogen, whereas the effluent contains the combustion products of the fuel used for reheating and thus, a higher heat capacity (cf. fig. 7.1, right). This, in turn, is accompanied by an increment of the mass flow density in flow direction, further increasing the heat flux capacity towards the reactor outlet.

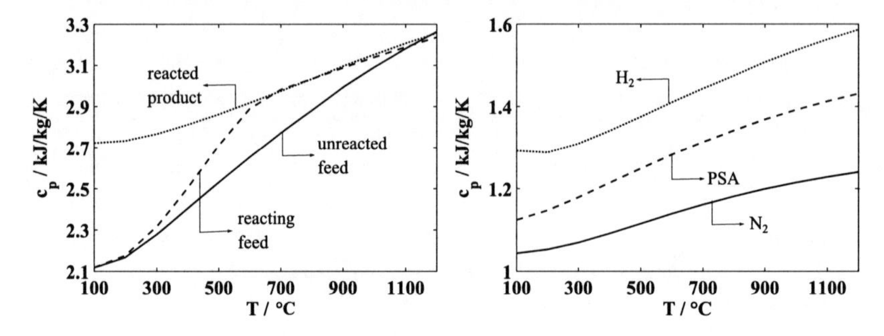

Figure 7.1: Left: heat capacity against temperature during production. CH_4 and H_2O gas mixture with constant composition ($S:C=3$) (straight line); the former mixture attains thermodynamic equilibrium conditions at each temperature (dashed line); gas mixture with constant composition corresponding to equilibrium at 800 °C (dotted line).
Right: heat capacity against temperature during regeneration. Heat capacity of a nitrogen flow (straight line); heat capacity of a flow containing combustion products of PSA off-gas diluted in N_2 (dashed line); heat capacity of a flow containing combustion products of H_2 diluted in N_2 (dotted line).

The effects on the reactor behaviour are well reflected in the evolution of temperature and conversion profiles during production plotted in figure 7.2. The left representations show the results obtained for operation with constant, equal heat capacity in both directions, whereas the right ones show those corresponding to the stationary state achieved when accounting for state dependency of the heat capacity. The straight lines correspond to the profiles at the beginning and end of the production period ($t = 0$ s and $t = 100$ s respectively), whereas the dotted lines represent profiles every 20 s. The behaviour with constant heat capacity can be very well predicted based on the fundamentals of the reactor concept (sec. 2.3.2). However, this is not the case when state dependency of heat capacity is accounted for.

In agreement with the analysis of figure 7.1, the significantly larger heat capacity of the flow during the production period causes a shift of the temperature and concentration profiles towards the right reactor outlet. Consequence thereof is a rapid decrease in conversion as soon as the high temperature region leaves the active packing. Besides this, regenerative heat exchange in the right reactor zone cannot be properly balanced, given the low heat capacity of the flow during the reheating period. Accordingly, the product flow cannot be efficiently cooled down and the right reactor outlet reaches unacceptable temperatures during operation.

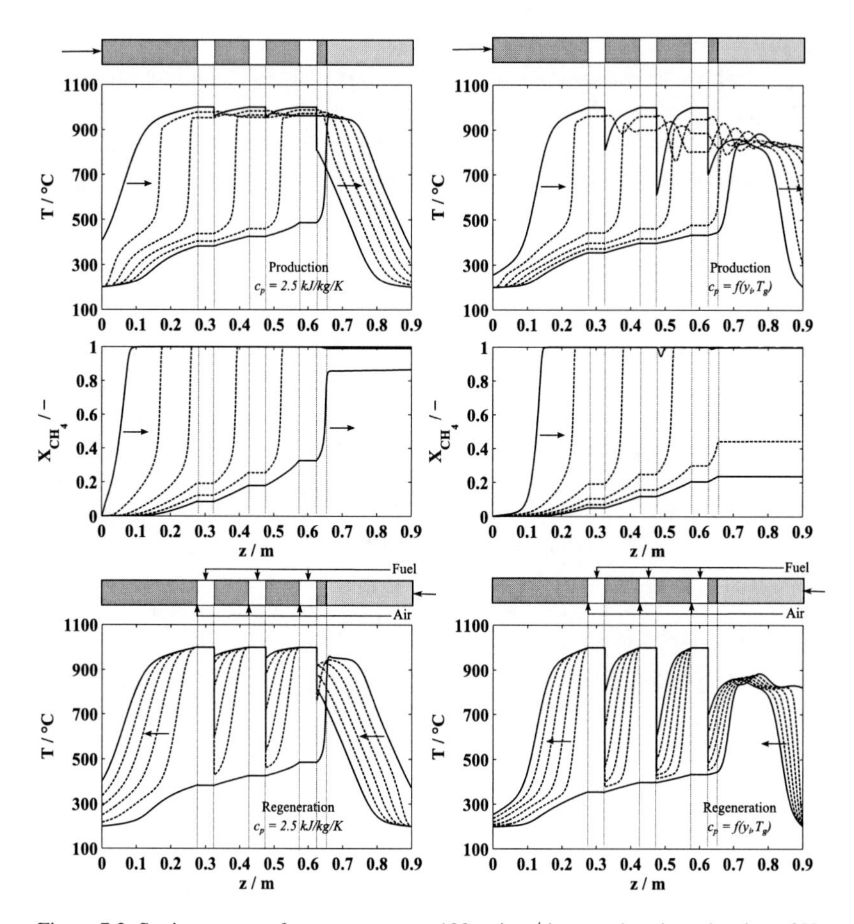

Figure 7.2: Stationary state for $\tau_{prod} = \tau_{reg} = 100\,\mathrm{s}$, $\dot{m}_{reg}/\dot{m}_{prod} = 1$ and combustion of H_2 for regeneration. From top to bottom: temperature and CH_4 conversion profiles during production and temperature profiles during regeneration. Profiles every 20 s. Left: constant heat capacity in both directions ($c_p = 2.5\,\mathrm{kJ/kg/K}$); right: $c_p = f(y_i, T_g)$.

A further consequence of the low heat capacity of the flow during regeneration is the insufficient convective energy transport from the combustion chambers into the packed bed (cf. figure 7.2, bottom, right). Despite the heat flux increase caused by the addition of fuel and combustion air in each of the chambers, it turns difficult to establish a uniform temperature profile at the end of the regeneration period, affecting the efficiency of the catalyst usage. Based on these observations, it becomes clear that the heat flux capacity balance in both directions can be improved through adjustment of the mass flow ratio between production and regeneration or the duration of the respective periods. The effect of such actions is discussed in the following.

7.1.2 Adjustment of operation parameters

The behaviour described in the previous section suggests that larger $\dot{m}_{reg}/\dot{m}_{prod}$ or τ_{reg}/τ_{prod} ratios can partially contribute in counteracting the heat flux asymmetry between both semi-cycles. Since alternating operation of two parallel reactors is striven, it is however advisable to keep the duration of the cycles unchanged and adjust the mass flow ratio instead[1]. The result of doing so can be discussed based on the temperature profiles under stationary state shown in figure 7.3 and the performance indicators summarized in figure 7.4.

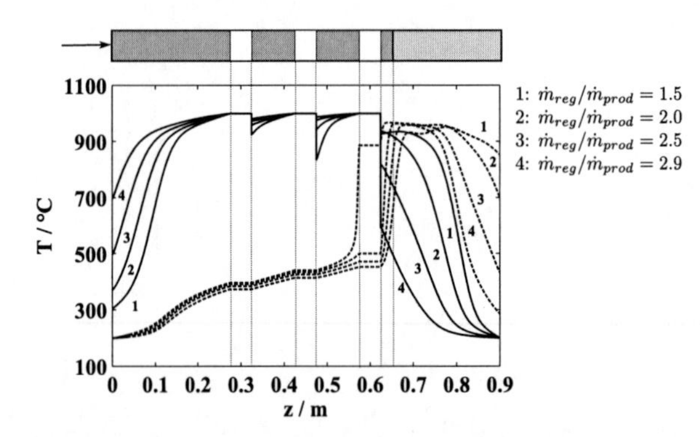

Figure 7.3: Influence of the mass flow ratio on the periodic stationary state reached for $\tau_{prod} = \tau_{reg} = 100\,\text{s}$ and H_2 as regeneration fuel. End of the production (dashed line) and the regeneration (straight line) periods.

[1]Notice that during the experimental work reported in chapter 5, the duration of the production and regeneration periods has also been varied. This is due to the fact that an arbitrary variation of the $\dot{m}_{reg}/\dot{m}_{prod}$ ratio in the experimental setup is limited.

Low $\dot{m}_{reg}/\dot{m}_{prod}$ ratios result in somewhat lower average temperature levels in the catalytic zone, provided that energy stored in the packing during regeneration is proportional to the flow rate (s. fig. 7.3). Accordingly, the endothermic reaction front travels further in flow direction during production and conversion rapidly decays as the front leaves the catalytic packing (cf. profiles 1 and 2 in figs. 7.3 and 7.4, c). Energy losses with the reactor effluent during regeneration are reduced (cf. fig. 7.4, a) but in contrast, the ones during the production period increase overproportionally (cf. fig. 7.4, b) since the flow fed from the right reactor end does not sufficiently cool down the inert packing.

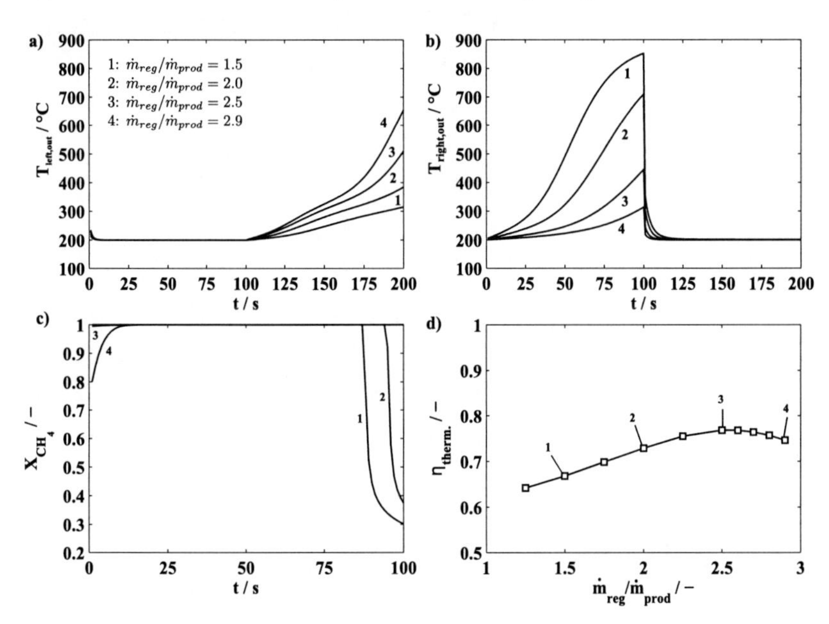

Figure 7.4: Temperature evolution at the a) leftmost and b) rightmost reactor end during a complete operation cycle. Conversion against production time (c) and thermal efficiency against mass flow ratio (d). Production: $0 \leq t < 100$; regeneration: $100 \leq t < 200$.

This behaviour is compensated when increasing the mass flow ratio. Energy introduced in the system during regeneration is larger and the usage of the catalyst packing becomes more efficient. Simultaneously, energy lost with the reactor effluents during regeneration increases and in turn, the oulet temperature in the rightmost reactor end can be significantly lowered. In summary, increasing the $\dot{m}_{reg}/\dot{m}_{prod}$ ratio contributes balancing the heat flux during both periods. Consequently, the rightmost regenerative heat exchange zone is more efficiently used and the catalyst usage efficiency can be improved. As a result thereof, larger productivities over the whole production period and hence, increased thermal efficiencies, can be achieved. The trade-off between heat losses through the reactor boundaries and the system productivity derives in an optimal operation range, which can be identified in a local maximum in the thermal efficiency curve for $2.5 < \dot{m}_{reg}/\dot{m}_{prod} < 2.7$ as depicted in figure 7.4, d.

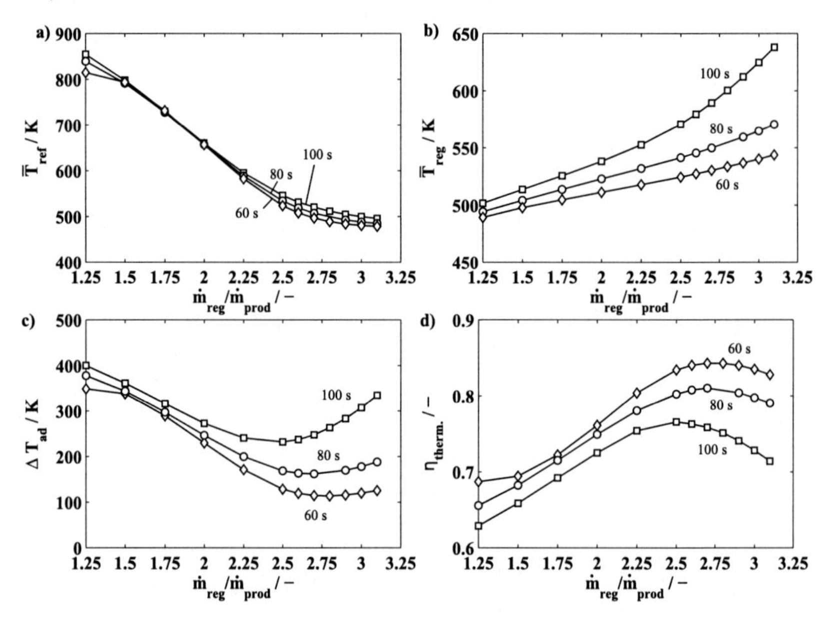

Figure 7.5: Performance of the RFR under periodic stationary state as a function of $\dot{m}_{reg}/\dot{m}_{prod}$ ratio and three switching frequencies: 60 s (diamonds), 80 s (circles) and 100 s (squares). Reheating temperature $T_{heat} = 1273.15\,\mathrm{K}$ and H_2 as fuel. a) and b) average effluent temperature during production and regeneration respectively; c) adiabatic temperature increase of the reactor effluents; d) thermal efficiency.

The behaviour depicted in representations a and b in figure 7.4 reveals that the cummulative energy lost with the product effluents increases overproportionally with the duration of a complete cycle. Accordingly, high switching frequencies contribute maximizing thermal efficiency during operation. This is summarized in figure 7.5, where the effect of reducing the duration of the endothermic and exothermic semicycles on the effluent temperatures and thus, on the energy losses and the thermal efficiency is represented against $\dot{m}_{reg}/\dot{m}_{prod}$.

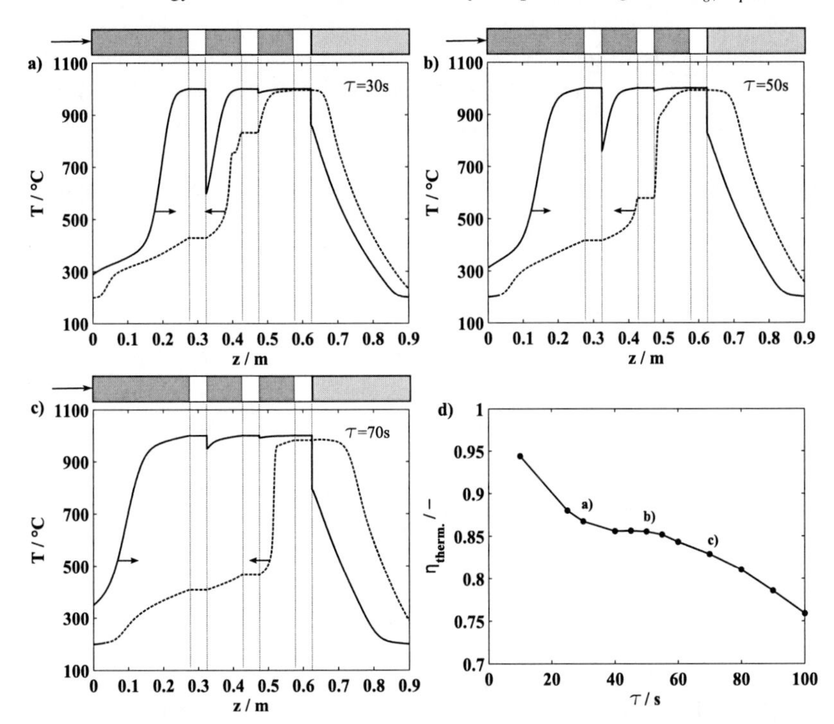

Figure 7.6: Performance of the RFR under periodic stationary state as function of the switching frequency for $\dot{m}_{reg}/\dot{m}_{prod} = 2.7$. Reheating temperature $T_{heat} = 1273.15\,\mathrm{K}$ and H_2 as fuel. a) to c) temperature profiles at the beginning of the production (—) and regeneration (– –) steps respectively; d) thermal efficiency against the switching frequency.

Simultaneously, the duration of the semicycles directly correlate with the distance travelled by the reaction and thermal waves during production and regeneration respectively. Hence, large switching frequencies increase the thermal efficiency of the system while the catalyst usage efficiency is reduced, since the extension of the zone where the endothermic and exothermic reactions are coupled diminishes. This behaviour is reflected in figure 7.6, for a constant mass flow ratio of $\dot{m}_{reg}/\dot{m}_{prod} = 2.7$.

For the given reactor dimensions and configuration, this ratio corresponds to that providing largest thermal efficiencies. Based on the temperature profiles in representations a to c, switching frequencies in the range between 50 and 90 s seem to be apropriate for an efficient catalyst usage at expense of a slight reduction of the thermal efficiency. In accordance to these results, figure 7.7 shows the thermal efficiency attained as a function of the average methane conversion at several switching frequencies. Whereas the curve for a switching frequency of $\tau_{prod} = \tau_{reg} = 100\,\mathrm{s}$ shows an optimum, lower switching frequencies enable operation with full methane conversion during the whole production period. The thermal efficiency, in turn, varies according to the operation parameters (i.e. $\dot{m}_{reg}/\dot{m}_{prod}$) and might reach values close to 85%, which emphasizes the excellent heat integration of the system. It is worth noticing that in practice, the reactor dimensions should be reconsidered in order to operate at technically practicable switching frequencies.

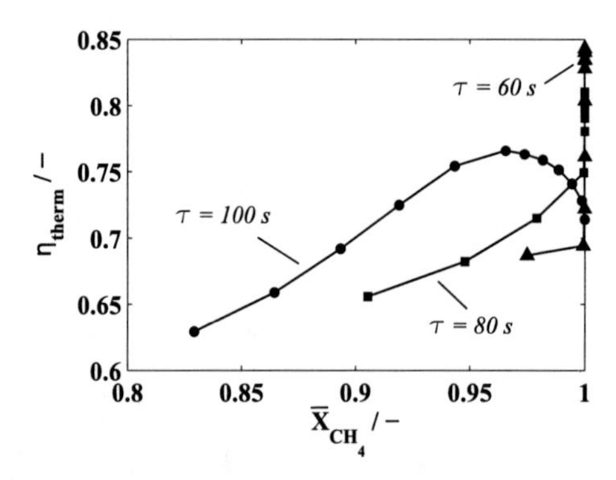

Figure 7.7: Thermal efficiency against average methane conversion at switching frequencies of $\tau_{prod} = \tau_{reg} = 60$, 80 and 100 s respectively. Reheating temperature $T_{heat} = 1273.15\,\mathrm{K}$ and H_2 as fuel.

In summary, the results discussed corroborate that an adjustment of the mass flow ratio $\dot{m}_{reg}/\dot{m}_{prod}$ and the switching frequency enables operation of the RFR ensuring an extended region for the coupling of the endothermic and exothermic reaction zones and exhibiting thermal efficiencies in the range of 80%. These efficiency values equal those estimated for the most efficient, large-scale MSR technologies with plant-wide heat integration (cf. section 2.2) and validate the RFR as a competitive technology for decentralized applications. Fine tuning of the operation parameters discussed above might be required in case that different fuels or reheating temperatures than those used in the current analysis are chosen.

These aspects, as well as the effect introduced by nonidealities in the adiabatic conditions of the system, are briefly reported within a parameter study summarized in appendix D.1.

7.2 Alternative reactor configurations for asymmetric operation

Unefficient usage of the catalyst represents an important drawback in the present reactor configuration. For instance, the temperature profiles in figure 7.6 a to c, reveal that the leftmost active zone only stores enough energy to ensure large methane conversions over 50 to 60% of its total length. This aspect has been considered and discussed by Glöckler et al. in [3], where the substitution of the leftmost reactor zone (or part of it) by inert packing is proposed. Using the leftmost packed bed section purely as a heat exchanger, the feed can be preheated during the endothermic period in a similar way to that taking place during the regeneration one. The result thereof is a significant increase of the temperature at the entrance of the catalytic zone. As described in section 2.3.1, increasing the latter beyond a given threshold ($\approx 775\,^{\circ}$C for the current case), the endothermic front occurring does not consist of a combined dispersive and a compressive wave, but only of a stationary and a travelling shock front as shown in the results of a detailed simulation in figure 7.8, a and b respectively.

Experimental proof of the behaviour depicted in figure 7.8 b has been previously reported in [3, 58, 59] and introduced with figure 2.5. Thus, its features are only briefly mentioned at this stage. Since the lowest temperature in the shock front is higher than that in a combined dispersive/compressive front (cf. figures a and b), the velocity of the front w_{shock} when a preheated flow enters the catalytic zone is lower (see equations (2.10), (2.12) and (2.15)). Accordingly, a more efficient usage of the catalyst for equal conversion rates and similar active bed length should be possible. Moreover, the mechanical stability of the packing can be enhanced, given that the operative temperature range is significantly reduced. In addition, the profiles obtained simplify the task of defining an appropriate structure for the packed bed.

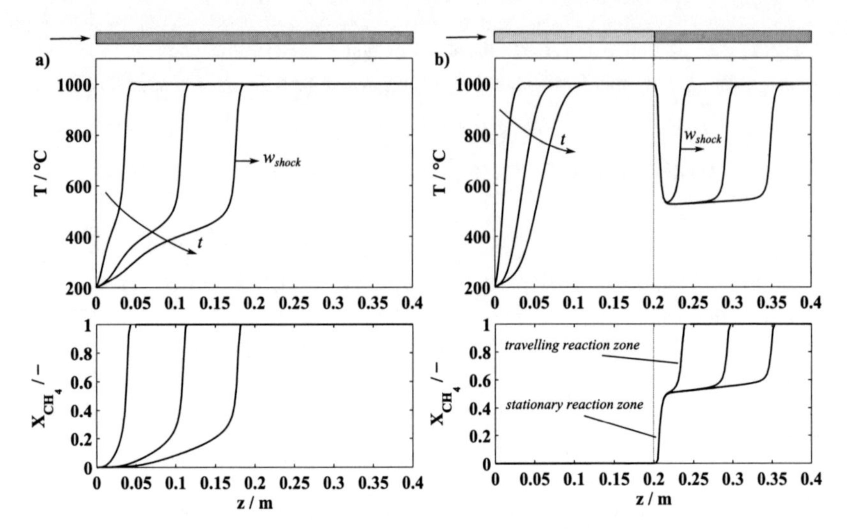

Figure 7.8: Comparison of operation with (left) and without (right) inert packing at the reactor inlet during production. Operating conditions: $T_{heat} = 1273.15\,\mathrm{K}$; $\dot{m}_{prod} = 1\,\mathrm{kg/m^2/s}$; $S:C = 3$.

In order to discuss the performance of reverse-flow operation taking advantage of the characteristics mentioned above, the leftmost and rightmost catalytic zones of the reactor design reported so far are replaced by inert packing. Mantaining the total length of the packed bed unchanged, this implies a reduction of the active length of almost 60%, from 0.475 to 0.2 m, located in the central part of the reactor. Accordingly, comparison with the operation performance reported for the former reactor concept requires the adjustment of the switching frequency, i.e. a 100 s semicycle in the concept discussed so far is equivalent to approximately 35 s in the reactor configuration with inert boundaries. A comparison of the thermal efficiency achieved with both systems against the $\dot{m}_{reg}/\dot{m}_{prod}$ ratio is depicted in figure 7.9, revealing a significant efficiency loss compared to that obtained with the original reactor structure under analogous operating conditions.

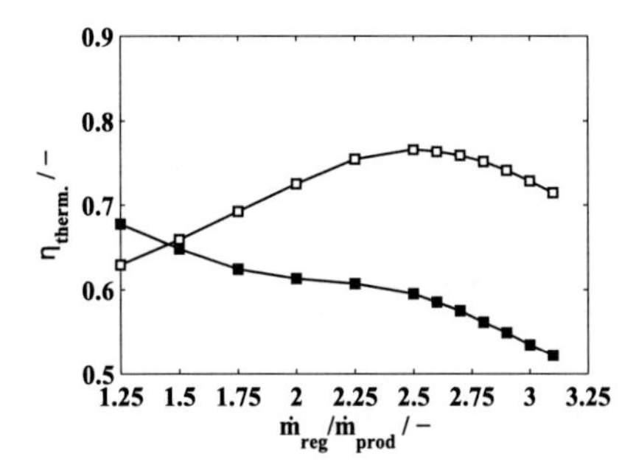

Figure 7.9: Energy efficiency comparison for operation with and without inert inlet section (black and white markers respectively). Operating conditions: $T_{heat} = 1273.15\,\mathrm{K}$; H_2 as fuel; $\tau = 35\,\mathrm{s}$ (black markers) and $100\,\mathrm{s}$ (white markers).

Despite the apparent attractiveness of the concept, the origin of the performance loss is discussed based on figure 7.10. The latter shows the stationary temperature and averaged heat flux profiles along the packed bed for $\dot{m}_{reg}/\dot{m}_{prod} = 1.25$ (left) and 2.5 (right). The first mass flow ratio corresponds to the simulated operation point with largest efficiency in the current configuration (s. fig. 7.9), whereas the second corresponds to the point exhibiting efficiency maxima with the former reactor design.

The temperature profiles in the upper part of figure 7.10 and the corresponding heat flux profiles in the lower part, reveal the difficulty to balance the convective energy flux in both directions of the flow. If the mass flow ratio is such that the heat fluxes at the leftmost inert section are balanced (figure a), energy is efficiently recovered in the latter but not in the rightmost zone and vice versa (figure b). In addition, according to the analysis reported in section 7.1.2, large $\dot{m}_{reg}/\dot{m}_{prod}$ ratios are preferred in order to improve catalyst usage efficiency, as reflected in the temperature profiles within the active reactor zone in figure 7.10. Equally, high productivity and full conversion can only be guaranteed at large mass flow ratios, provided that the energy stored in the packed bed must be large enough to keep the reaction front within the active zone during the whole production period. Unfortunately, the effect of effluent heat losses is especially significant for large $\dot{m}_{reg}/\dot{m}_{prod}$ ratios, since the effluent temperature is considerably larger than in the opposite case. Interestingly, this drawback was not meaningful in the original reactor configuration. The reason therefor is that the energy surplus supplied to the leftmost packing zone can be consumed during the subsequent production period, whereas in the current configuration, energy accumulates in the inert region and the excess leaves the system with an effluent exhibiting prohibitively high temperatures.

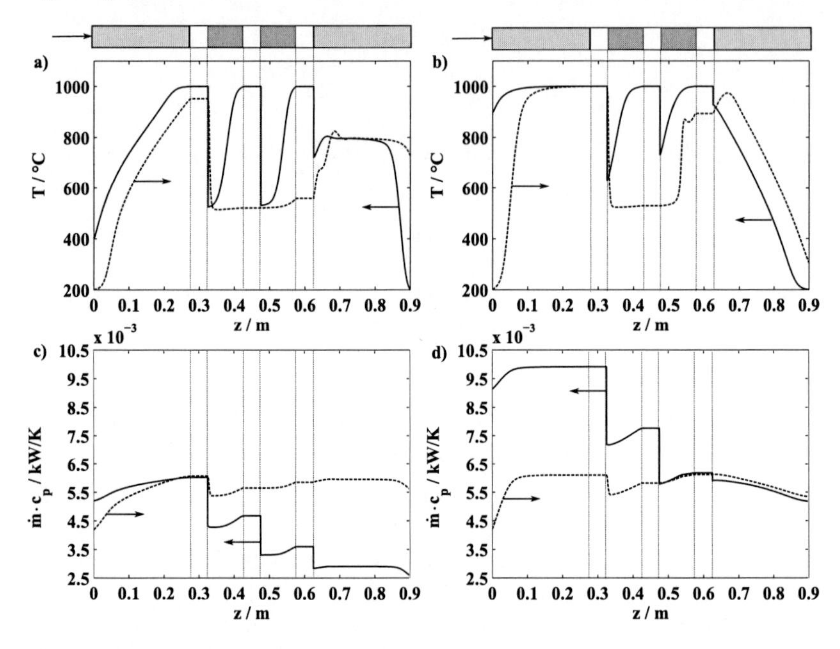

Figure 7.10: Steady state operation for $\tau_{prod} = \tau_{reg} = 35\,\mathrm{s}$ and $\dot{m}_{reg}/\dot{m}_{prod} = 1.25$ (figures a and c) and 2.5 (figures b and d).
Top: Temperature profiles at the beginning and end of the production period (straight and dashed lines respectively).
Bottom: Average heat flux during production (dashed lines) and regeneration (straight lines).

An approach to improve the performance of the current reactor configuration consists of balancing the heat flux in the rightmost reactor zone and mantaining the latter constant over the reactor length. A possible scenario to achieve this goal is to keep the mass flow along the reactor during regeneration unchanged. This is thinkable for larger scale RFR, which could potentially use conventional FLOX® combustor technology instead of the combustion chambers used in the current study. The former are designed in order to drain a continuous flow of exhaust gases, which is further used to preheat the air and fuel flows via recuperative or regenerative heat exchange [63]. As summarized in figure 7.11 a, drainage from each chamber of the mass being fed for combustion proves to effectively balance the heat flux along the reactor length.

This adjustment of the convective fluxes in both directions translates in a noteworthy reduction of the average adiabatic temperature increase of the reactor effluents during operation (figure b), significantly enhancing the operability of the system. In spite thereof, a significant amount of the energy released during regeneration is systematically withdrawn with the mass flow purged and does not contribute restoring the energy stored in the packed bed. Consequently, the thermal efficiency of the system remains below that obtained with the former reactor configuration under analogous conditions (cf. fig. 7.11, d).

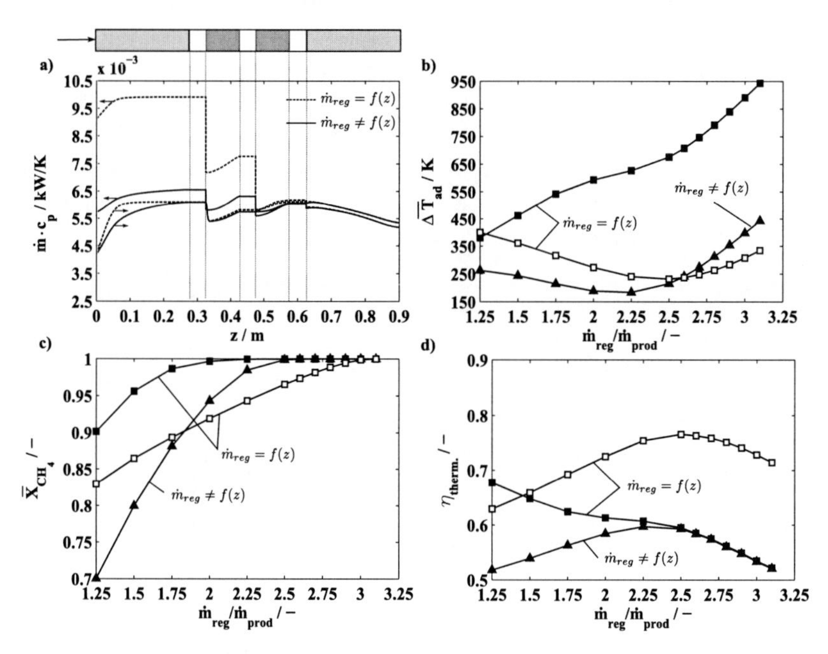

Figure 7.11: Reactor configuration: inert boundaries and $\tau = 35\,\mathrm{s}$ (black markers); active left zone and $\tau = 100\,\mathrm{s}$ (white markers).
a) Average heat flux over the reactor length during production (→) and regeneration (←) for $\dot{m}_{reg}/\dot{m}_{prod} = 2.5$.
b) Average effluent adiabatic temperature increase.
c) Average methane conversion during production period.
d) Thermal efficiency.

The energy being drained can be optionally used in downstream units, e.g. to preheat the fuel and combustion air being fed to the system during regeneration. A best-case estimation of the resulting thermal efficiency can be made under the assumption that the fuel and air flows can be ideally preheated to the desired reheating temperature (i.e. 1000 °C) with the energy withdrawn during regeneration. In this case, 20 to 25% larger thermal efficiencies than those shown in figure 7.11, d (triangles as markers) could be reached. Accordingly, the performance of the system would become comparable to that obtained with the reactor configuration described in section 7.1.

If only energy integration within the reactor is considered, figure 7.12 shows a comparison between the thermal efficiency against the average methane conversion for operation with and without adjustment of the heat fluxes during production and regeneration ($\dot{m}_{reg} \neq f(z)$ and $\dot{m}_{reg} = f(z)$, respectively). This representation of the results shows that the overall performance of the system consists of a trade-off between thermal efficiency and productivity and lays below that obtained with the original RFR configuration (cf. figure 7.7). Alternatively, section 7.3 proposes a further operation scheme for the current reactor configuration. With this operation scheme, some limitations and in particular those affecting the operability of the system, can be partly overcome.

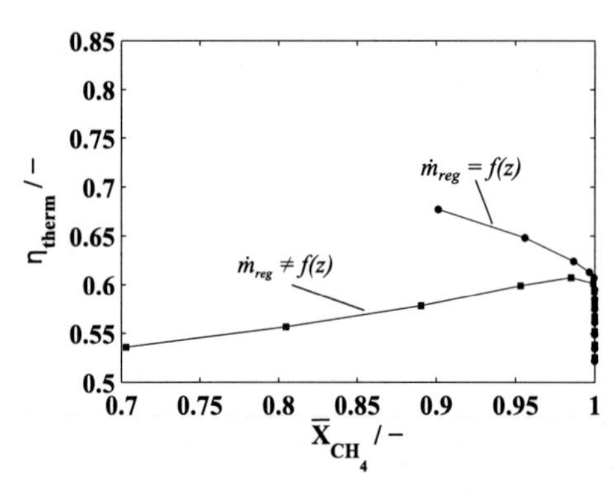

Figure 7.12: Thermal efficiency against average methane conversion for $\tau_{prod} = \tau_{reg} = 35\,\text{s}$. Reheating temperature $T_{heat} = 1273.15\,\text{K}$ and H_2 as fuel.

7.3 Symmetric operation

The attractivity of the reactor configuration discussed in the previous section relies mainly in the efficient catalyst usage and in particular, in the use of active material only in those reactor zones contributing to the productivity of the system. Energetically, however, the concept presents some drawbacks in comparison to the former reactor configuration described in section 7.1. In spite thereof, the current distribution of active and inert sections in the packed bed enables exploring alternative operation patterns to those described so far.

This section is dedicated to the preliminary analysis of a reverse-flow reformer operating in a pseudo-symmetric mode. This means that production and regeneration steps are further sequentially run, whereas the direction of the flow is only reversed after a complete cycle, comprising both steps, is finished. This switching pattern was already proposed and thoroughly analysed by Van Sint Annaland for the sequential coupling of the endothermic propane dehydrogenation and the combustion of methane [4, 35], provided that it inherently fulfills the condition of zero differential creep velocity. Moreover, he proved that the switching scheme with the pattern $\overset{prod}{\rightarrow}\overset{reg}{\rightarrow}\overset{prod}{\leftarrow}\overset{reg}{\leftarrow}$ exhibited the largest energy efficiencies [4].

The reactor system in the current study, in turn, can additionally benefit from the switching strategy, given that the reactor design can be significally simplified. The leftmost catalytic reactor end can be replaced with inert packing, resulting in a fully symmetric reactor configuration. Moreover, the operation principle might enable the suppression of the leftmost and rightmost combustion chambers, reducing the design complexity additionally.

The ideal operation principle can be represented according to the sketches in figure 7.13, where the packed bed is initially preheated to a desired temperature level corresponding to the plateau in the middle of the reactor. As in the previously discussed concepts, temperature decreases towards the reactor inlet and outlet, so that efficiency by thermal regeneration can be guaranteed. Given this initial temperature profile, the reactor can be supplied with the reforming reactants from the left reactor end (fig. 7.13, top, left). The formation of both a stationary reaction zone at the entrance of the catalytic packing and a moving reaction front travelling in the flow direction resembles the behaviour last discussed in the previous section. In the current concept, however, production is stopped as soon as the shock wave reaches the combustion chamber located in the middle of the reactor ($t_1 = t_0 + \tau_{prod}$). In this case two premises should be simultaneously fulfilled:

- the temperature at the rightmost transition between catalytic and inert packing is kept at a high level during the production period and thus, high conversion levels can be mantained, and
- the energy contained in the leftmost inert zone is sufficient to ensure high temperatures at the entrance of the catalytic zone during the regeneration period.

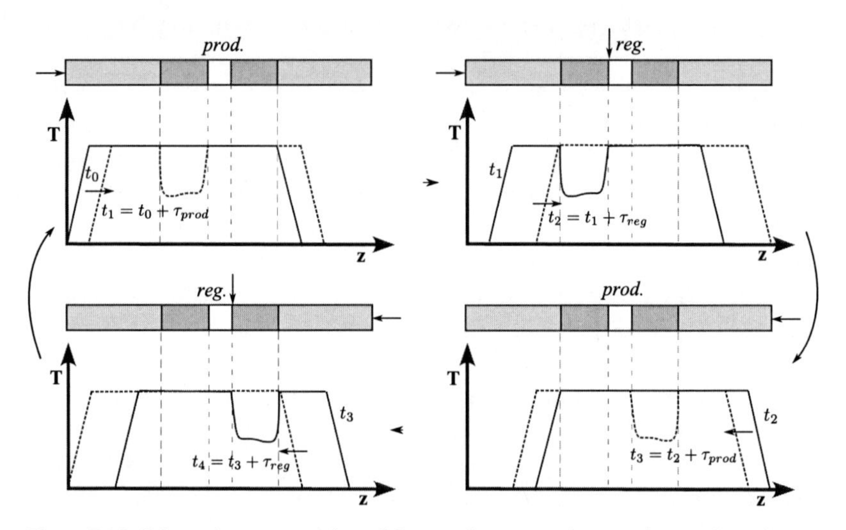

Figure 7.13: Schematic representation of the pseudo-symmetric operation mode under periodic stationarity. The upper sketches, left and right represent an appropriate initial state, at which a production step in the right direction is started and the subsequent regeneration step in the same direction, respectively. Sketches on the bottom represent a production (right) and regeneration (left) steps in the opposite direction.

As the end of the production step is reached, the regeneration step is started without reversing the direction of the flow (s. figure 7.13, top, right). Hence, an inert flow is supplied through the left reactor end and the energy contained in the inert packing is convectively transported into the catalytic zone that has been previously cooled down through the endothermic reaction. Simultaneously, the cold gases entering the combustion chamber located at the center of the reactor are heated up based on the same operation principle as in the previous concepts. The temperature of the gases leaving the combustion chamber corresponds to the temperature of the initial temperature plateau. As soon as the leftmost catalytic zone is reheated ($t_2 = t_1 + \tau_{reg}$), the regeneration step can be finished and the flow direction can be reversed. As shown in figure 7.13, bottom, the production (right) and regeneration (left) steps are performed until the initial temperature profile (at t_0) is established, closing the cycle of a periodic stationary operation mode.

7.3.1 Simulation results

The analysis of the symmetric operation pattern is performed with a reactor configuration resembling the one used in the previous section. The total length of the packed bed is mantained as in the configurations discussed so far (cf. sections 7.1 and 7.2), whereas only the central combustion chamber is preserved. The chamber represents the symmetry axis, accounting for 375 mm of packing at each of its sides. Each of them consists of 100 mm active packing and 275 mm long inert sections serving as regenerative heat exchangers at both reactor ends. The duration of the production step is adjusted to the distance traveled by the endothermic front, which is the equivalent of half of the active length[2]. Accordingly, production should not exceed 20 s.

Figure 7.14 shows the evolution of simulated temperature and conversion profiles under steady state operation. The production period (top, left) extends over 20 s, whereas the regeneration period does it over 40 s (top, right), taking into account that the reaction front travels approximately twice as fast as the thermal one (cf. sec. 2.3.2). Additionally, in order to balance the heat capacity difference between both steps, the mass flow ratio $\dot{m}_{reg}/\dot{m}_{prod}$ is set to 1.7. The temperature profile marked as t_2 in figure 7.14 b, corresponds to the initial temperature profile of the subsequent production period, once the flow direction is reversed.

At first sight, the divergence between the temperature profiles in figures 7.13 and 7.14 becomes obvious. The impossibility to mantain the form stability of the thermal front due to the dispersive character of the latter, enhanced by the effects of the state dependant heat capacity of the flow, constitutes a major drawback in the realisation of the concept. The profiles at the reactor boundaries are significantly flattened, deteriorating the thermal efficiency of the system.

A close analysis of plots a and c in figure 7.14 reveals the formation of the stationary reaction zone at the entrance of the active section and that of a travelling one, which moves towards the combustion chamber. Whereas full conversion is achieved before entering the

[2]It should be noted that the strategy followed for the comparison of the different reactor concepts implies the preservation of the original reactor dimensions and operation flows, whereas the switching frequency is adapted to the length of the catalytic packing. Larger semicycles might be attained by appropriate scale-up of the reactor dimensions.

chamber, the low temperature level at the outlet of the active zone favors back reaction and the effective conversion remains significantly below 90 % over the production period. In order to counteract this behaviour and enhance productivity, larger temperatures at the outlet transition between active and inert packing in the flow direction should be mantained during operation. Therefor, larger energy amounts are to be supplied and convectively transported towards the reactor outlet during the regeneration step. Hence, longer regeneration steps or alternatively, larger $\dot{m}_{reg}/\dot{m}_{prod}$ ratios are required.

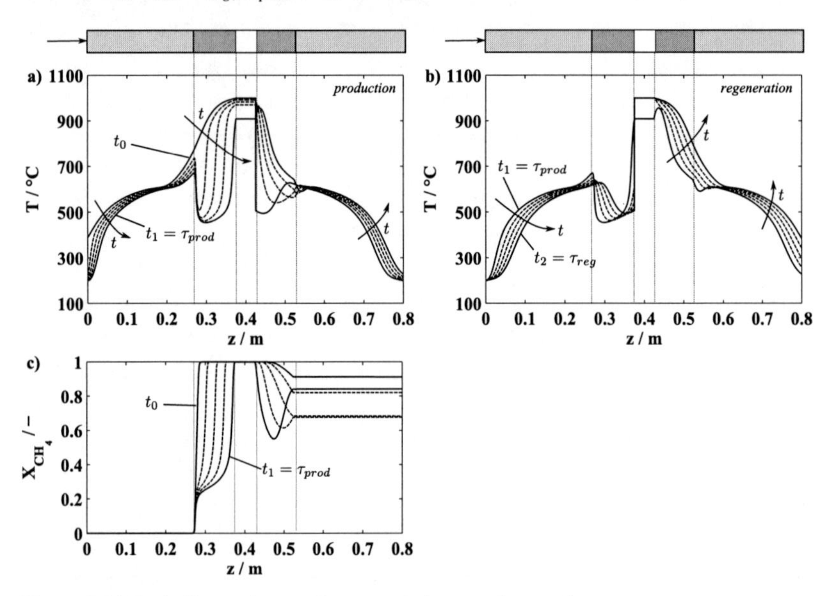

Figure 7.14: Periodic steady state for symmetric operation mode.
a) Production step: duration $\tau_{prod} = 20\,s$; temperature profiles every 5 s.
b) Regeneration step: duration $\tau_{reg} = 40\,s$; temperature profiles every 10 s; H_2 used as fuel; $T_{heat} = 1273.15\,K$; $\dot{m}_{reg}/\dot{m}_{prod} = 1.7$.
c) Methane conversion profiles corresponding to the temperature profiles plotted in figure a).

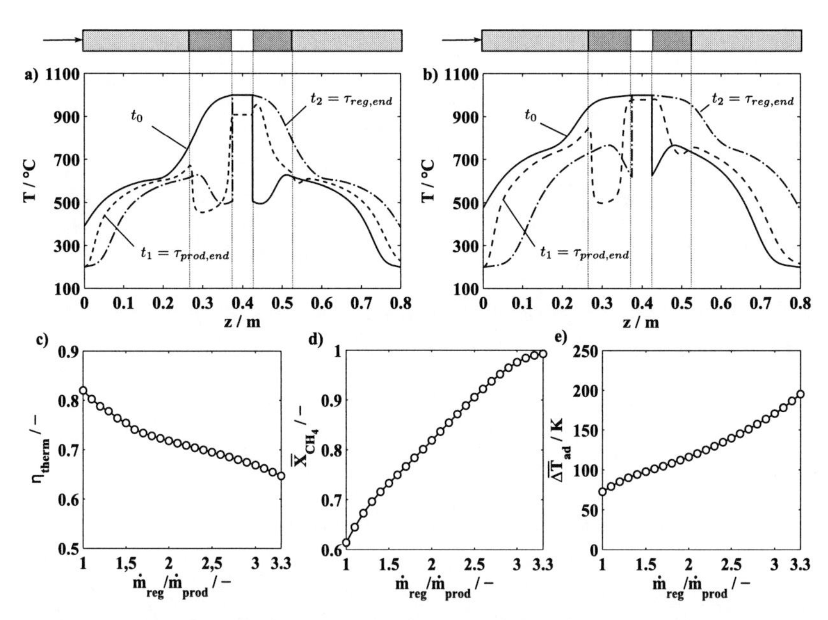

Figure 7.15: Top: temperature profiles at the beginning (straight line) and at the end (dashed line) of the production step, as well as at the end of the regeneration step (dash-pointed line) in one flow direction. $\tau_{prod} = 20$ and $\tau_{reg} = 40$ s for a) $\dot{m}_{reg}/\dot{m}_{prod} = 1.7$ and b) $\dot{m}_{reg}/\dot{m}_{prod} = 3.3$ respectively.
Bottom: performance indicators for steady state operation against the $\dot{m}_{reg}/\dot{m}_{prod}$ ratio. c) Thermal efficiency; d) average methane conversion; e) average effluent adiabatic temperature increase.

The effect of doubling the mass flow ratio in comparison to the reference configuration is summarized in the upper plots of figure 7.15. The bottom plots, in turn, illustrate relevant performance indicators against a wide $\dot{m}_{reg}/\dot{m}_{prod}$ range. The performance indicators exhibit an analogous behaviour to that observed for operation under asymmetric conditions for a reactor configuration with inert boundaries (cf. fig. 7.9), in which an increase in productivity directly correlates with a diminishing thermal efficiency due to augmented heat losses. Thus, contrary to the original asymmetric operation mode discussed in section 7.1, which exhibits a performance optimum resulting from the trade-off between productivity and energy losses, the current operation mode is characterized by directly competing average methane conversion and thermal efficiency.

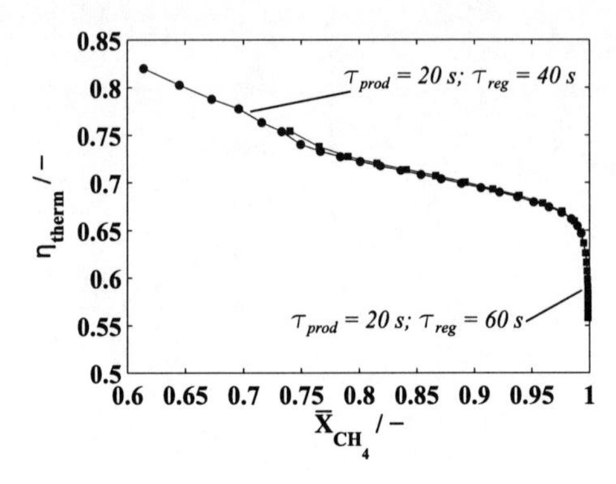

Figure 7.16: Thermal efficiency against average methane conversion for $\tau_{prod} = 20\,\mathrm{s}$ and $\tau_{reg}/\tau_{prod} = 2$ and 3 (black circles and squares respectively). Reheating temperature $T_{heat} = 1273.15\,\mathrm{K}$ and H_2 as fuel.

This behaviour can be further seen in figure 7.16, where the decreasing trend of the thermal efficiency at increasing productivities is shown. Interestingly, the overall performance trend remains almost unchanged if further operation parameters, such as the duration of the regeneration step, are changed. In this case, the choice of the proper operating parameters (i.e. $\dot{m}_{reg}/\dot{m}_{prod}$ and τ_{reg}/τ_{prod}) might be made considering the operability of the system by reducing the effluent exit temperatures (cf. appendix D.2).

The simplicity of the concept and the fact that it energetically outperforms the concept with inert reactor ends and asymmetric operation proposed by Glöckler et al. in [3] and discussed in section 7.2, encourages the examination of technical solutions to contribute rising the energetic integration degree of the system. Possible approaches are introduced in the concluding section of this chapter.

7.4 Summary and outlook

A direct comparison of the overall performance of the three reactor concepts discussed so far is possible based on figure 7.17, where representative thermal efficiency curves as a function of the methane conversion for each of the reactor configurations and operation modes are shown.

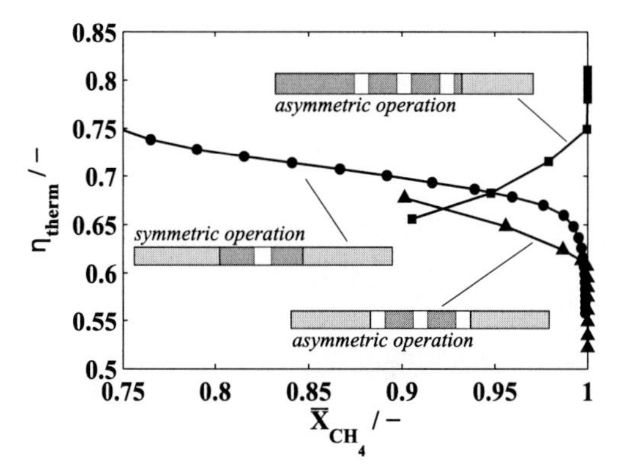

Figure 7.17: Representative thermal efficiency curves against methane conversion for three reactor configurations and operation modes:
Asymmetric reactor and operation (squares): $\tau_{prod} = \tau_{reg} = 80\,s$.
Symmetric reactor and asymmetric operation (triangles): $\tau_{prod} = \tau_{reg} = 35\,s$.
Symmetric reactor and operation (circles): $\tau_{prod} = 20\,s$ and $\tau_{reg} = 60\,s$.
Reheating temperature $T_{heat} = 1273.15\,K$ and H_2 as fuel.

The asymmetric RFR concept described in section 7.1 exhibits the largest energetic efficiencies from all concepts reported. The thermal efficiencies estimated based on the detailed simulations reported are in the same range than those achieved in the optimized state-of-the-art reforming technologies. Thus, it can be stated that the asymmetric RFR concept is especially suited for the production of hydrogen in decentralized facilities. Despite this, unefficient catalyst use remains a drawback, which cannot be easily solved by means of appropriate adjustment of the operation parameters. The catalyst usage can be improved with the concept proposed by Glöckler et al. in [3] and discussed in section 7.2, consisting on providing the reactor with inert ends. This approach, however, intensifies the difficulty to balance the heat flux capacities during the production and regeneration cycles, resulting in significantly lower energetic performances in comparison to the former RFR concept for comparable productivities.

Besides this, providing the RFR with two inert boundaries results in a symmetric reactor design, which opens the possibility to consider alternative operation patterns than those pursued so far. Within this context, running the RFR under symmetric operating conditions has turned to be an interesting alternative. As well as enabling a considerable simplification of the reactor design (only one combustion chamber is required), the overall performance of the reactor with inert ends is larger if operated symmetrically than in an asymmetrical mode (cf. figure 7.17). Based on these considerations, experimental validation of the concept is warmly encouraged.

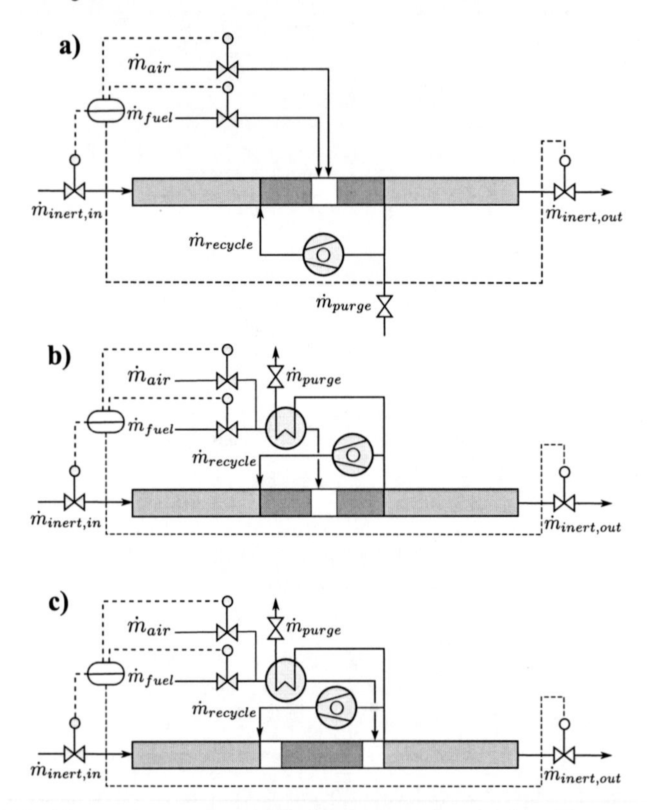

Figure 7.18: Scheme of a reactor conceived for symmetric operation. The reactor is provided with the possibility to purge and/or recycle part of the mass flow in the inner reactor zone, contributing to increase the convective transport in it. Heat fluxes at the reactor boundaries are reduced in order to minimize energy losses with the reactor effluents.

Given the simplicity of the reactor concept, the attractivity of the symmetric operation mode could be enhanced if a high temperature level in the interfaces between the inert and the catalytic packing can be maintained and, at the same time, operability is improved by effectively reducing the temperature of the reactor effluents. Therefor, operation limitations posed by the lack of form stability of the thermal fronts and the intrinsic heat flux capacity increase along the reactor axis in flow direction should be overcome. An approach to do so is sketched in figure 7.18 a, which makes use of the idea of purging and recycling part or the totality of the additional mass flow supplied to the system as fuel and combustion air during the regeneration period, e.g. with help of a turbocharger.

The concept sketched in figure 7.18 a, enables recirculation of larger mass flows in the inner zone of the reactor than those flowing through the boundaries. Thus, higher heat fluxes could be generated in the regions where a fast convective transport of the energy is desired, whereas the heat flux at the reactor boundaries, responsible for the energy losses with the effluents, could be minimized. A further positive aspect of the concept sketched is the contribution to keep a high temperature at the entrance of the leftmost catalytic zone if the recirculated flow $\dot{m}_{recycle}$ exhibits higher temperatures than the incoming flow $\dot{m}_{inert,in}$. This would be the case, for example, if $\dot{m}_{recycle}$ is directly drained from the combustion chamber at temperatures corresponding to the reheating temperature. In that case, however, the apparative realisation of the handling concept of flows exhibiting extremely high temperatures should be thoroughly considered.

Should an internal flow recirculation enhance the performance of the symmetric operation mode, extended concepts accounting for heat recovery of the purge leaving the system could be further analyzed (cf. figure 7.18, b). Moreover, internal flow recirculation could be applied to a reactor with a similar design to that described in section 7.2. As sketched in figure 7.18 c, placing at least two combustion chambers at both ends of the active section would make possible to run the reactor either in a symmetric or in an asymmetric mode. Hence, an estimation of the potential of such concepts based on detailed simulation studies is herewith strongly recommended.

Appendix A

Complementary information and experimental parameter sets

The information contained in this chapter provides complementary information to the results reported in chapters 3, 4 and 5. Experimental and simulation parameter sets used for the design of the combustion chamber as well as for its operation at understoichiometric and stoichiometric conditions are summarized. Additionally, relevant operation parameters of the RFR under periodic operation are reported. Self-explanatory captions are used.

A.1 Combustion chamber design

A.1.1 Experimental studies

General information

Table A.1: Representative flows for the design of the combustion chamber in relation to the reforming load during the production period. An energy release surplus of approx. 30% is accounted for, in order to balance heat losses in the laboratory reactor.

Reforming load $\left(\frac{kW_{LHV,H_2}}{l_{Reactor}}\right)$	$Q_{reforming}$ (kW)	$Q_{reheating}$ (kW)	$\dot{V}_{PSA}$ $\left(\frac{Nl}{min}\right)$	$\dot{V}_{inert}$ $\left(\frac{Nl}{min}\right)$	$\dot{V}_{air}$ $\left(\frac{Nl}{min}\right)$
0.50	0.25	0.34	2.02	3.03	4.75
2.00	1.01	1.34	8.10	12.25	19.00
5.00	2.52	3.36	20.20	30.30	47.50

Fuel characteristics

Equivalent energy input with H_2 and PSA off-gas as fuel require different amounts of fuel and combustion air. The corresponding values are summarized in the following table.

Table A.2: Energy density and required amounts of combustion air for operation with H_2 and PSA off-gas as fuel.

Fuel	Energy density (kJ/kg_{fuel})	Air required for stoich. combustion (kg_{air}/kg_{fuel})	Air required per energy input (kg_{air}/kJ)
H_2	$\approx$142	34.5	0.243
PSA off-gas	$\approx$10.77	2.95	0.274

Tracer methods in the water channel

Table A.3: Parameter sets for operation of the water channel (s. section 3.3.2). *MS* stays for *main stream* and *SS* for *side stream*. The Reynolds number is calculated at the outlet of one nozzle, in a $3x1\,mm$ nozzle head.

Air @ 475°C		Water @ 25°C			
MS	SS	MS	SS	$\frac{\dot{V}_{SS}}{\dot{V}_{MS}}$	Re_{SS}
$(\frac{Nl}{min})$	$(\frac{Nl}{min})$	$(\frac{l}{min})$	$(\frac{l}{min})$	-	-
101.1	24.91	2.07	0.8	0.386	6.38E+03
20.4	16.48	0.59	0.53	0.898	4.22E+03
20.4	5.44	0.59	0.18	0.305	1.39E+03
7.53	8.82	0.14	0.28	2	2.26E+03
7.53	3.61	0.14	0.12	0.857	9.24E+02

A.1.2 Temperature distribution in the chamber

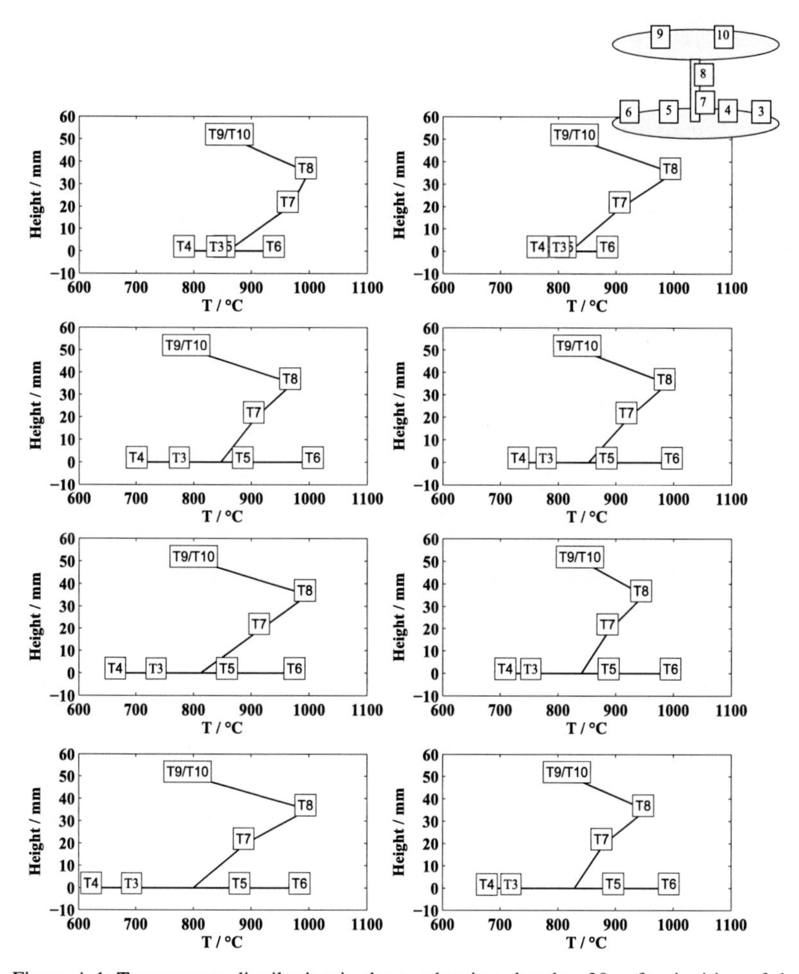

Figure A.1: Temperature distribution in the combustion chamber 30 s after ignition of the gas bulk for H_2 combustion (cf. sec. 4.2.2). Left figures: $T_{in} = 500\,°C$; right figures: $T_{in} = 550\,°C$. From top to bottom $\dot{V}_{inert} = 10$, 20, 30 and 40 Nl/min. The insert at the top right corner indicates the positions of the thermocouples T3 to T10 in the combustion chamber.

A.1.3 Simulation studies

The following table contains the parameters used to simulate the formation of flame fronts. As reference, the experimental operation point depicted in figure 3.12, left, and summarized as *Case 1* is taken. Cases *2* and *3* are chosen to emulate the experimental point using H_2 instead of PSA off-gas as fuel.

Table A.4: Case 1: Experimental operation parameters for PSA off-gas combustion, under fuel rich conditions and flame formation (cf. figure 3.12, left).
Case 2: Adaptation of the parameters to simulate case 1 using hydrogen as fuel.
Case 3: Adaptation of the parameters to simulate case 2 under stoichiometric conditions and similar main flow rates.

Case	**Flame**	$\dot{V}_{N_2}$ $\left(\frac{Nl}{min}\right)$	$\dot{V}_{fuel}$ $\left(\frac{Nl}{min}\right)$	$\dot{V}_{air}$ $\left(\frac{Nl}{min}\right)$	λ -	$T_{max,sim}$ °C	$\dot{V}_{N_2}/\dot{V}_{fuel}$ -
1	yes	15	10 (PSA)	7	0.3	–	1.5
2	yes	15	9.2 (H_2)	7	0.32	1536	1.6
3	no	30	3.3 (H_2)	7.8	1	1246	9.1

The last column in table A.4 provides the ratio between the inert flow fed with the main stream $\dot{V}_{N_2}$ and the fuel supplied to the chamber $\dot{V}_{fuel}$. The large inert-to-fuel ratio in case 3 is the main responsible for the large dilution effect of the reacting gases and hence, for flame suppression. This favourable effect enables operation under flameless conditions for significantly lower recirculation ratios than formerly assumed and reinforces the advantages of running the chamber under stoichiometric conditions. It can be stated, that recirculation ratios in the range between 1 and 2 are sufficient to suppress flame formation, whenever the mixing quality of the fuel in the chamber is good enough.

A.1.4 Flow rate variation and effect on the real factor

The combustion chamber can be described as a perfectly mixed gas phase reactor. Under dismisal of the term for the specific kinetic energy of the fluid and the work generated by volume changes, the energy balance can be written as follows

$$V_R \cdot \rho \cdot c_p \cdot \frac{\partial T}{\partial t} = \rho^+ \cdot {c_p}^+ \cdot \dot{V}^+ \cdot T^+ - \rho \cdot c_p \cdot \dot{V} \cdot T + \dot{Q}_c + V_R \cdot r \cdot (-\Delta h_R) \tag{A.1}$$

with the cooling capacity

$$\dot{Q}_c = U_W \cdot A_W \cdot (T_c - T), \tag{A.2}$$

where A_W represents the surface of the chamber exchanging heat with the environment $\left[m^2\right]$, U_W the heat transfer coefficient $\left[\frac{kW}{m^2 \cdot K}\right]$ and T_c the temperature at the reactor outer wall $[K]$.

In order to simplify the derivation, the system is considered to be operated under stationary state. Furthermore, inlet and outlet densities $(\rho = \rho^+)$, heat capacities $(c_p = {c_p}^+)$ and volume flows $\left(\dot{V} = \dot{V}^+\right)$ are assumed to be identical.

Considering the definition of the residence time τ

$$\tau = \frac{V_R}{\dot{V}} \Leftrightarrow \dot{V} = \frac{V_R}{\tau} \tag{A.3}$$

equation (A.1) can be written in a simplified form

$$\underbrace{\tau \cdot r \cdot (-\Delta h_R)}_{a} + \rho \cdot c_p \cdot (T^+ - T) = \tau \cdot \frac{U_W \cdot A_W}{V_R} \cdot (T - T_c) \tag{A.4}$$

The reaction rate r in term a can be substituted based on the general mass balance of a component j

$$\frac{dc_j}{dt} \cdot V_R = \dot{V}^+ \cdot {c_j}^+ - \dot{V} \cdot c_j + V_R \cdot \sum_{i=1}^{I} \nu_{ij} \cdot r_i \tag{A.5}$$

Based on the previous assumptions and setting the mass balance as stationary, for a component A that reacts in the form $A + B \rightarrow C$ (e.g. hydrogen combustion), the term r_A can be written as a function of the the inlet concentration of component A, c_A^+, and the conversion in the system X_A

$$r_A = \frac{\dot{V}}{V_R} \cdot {c_A}^+ \cdot X_A \tag{A.6}$$

Introducing equation (A.6) in equation (A.4) results in following expression

$$\underbrace{\tau \cdot \frac{U_W \cdot A_W}{V_R} \cdot (T - T_c)}_{\mathbf{a}[\frac{\mathrm{kJ}}{\mathrm{m}^3}]} = \underbrace{{c_A}^+ \cdot X_A \cdot (-\Delta h_R)}_{\mathbf{b}[\frac{\mathrm{kJ}}{\mathrm{m}^3}]} + \underbrace{\rho \cdot c_p \cdot (T^+ - T)}_{\mathbf{c}[\frac{\mathrm{kJ}}{\mathrm{m}^3}]} \tag{A.7}$$

Setting the heat transfer coeffcient U_W, conversion of component A and even the temperature gradients $(T - T_c)$ and $(T^+ - T)$ in terms a and c respectively as constant, variations in the volume flow rates supplied to the chamber do only affect the specific cooling term a and the reaction term b. Besides changes in the inert volume flow $\dot{V}_{inert}$, variations in the so called *real factor RF* also have an influence on the paramters τ and c_A^+ (cf. sec. 4.2.2).

The inlet concentration of component *A* can be written as follows

$$c_A^+ = \frac{RF \cdot \dot{N}_A^+}{\dot{V}} \tag{A.8}$$

with $\dot{N}_A^+$ being directly proportional to $\dot{V}_{inert}$

$$\dot{N}_A^+ \propto \dot{V}_{inert}$$

The denominator of equation (A.8) accounts for all the streams being fed to the combustion chamber

$$\dot{V} = \sum_{i=1}^{I} \dot{V}_i^+ \tag{A.9}$$

$$\dot{V} = \dot{V}_{inert}^+ + RF \cdot \left(\dot{V}_{fuel}^+ + \dot{V}_{air}^+\right)$$

At the same time, $\dot{V}_{fuel}^+$ and $\dot{V}_{air}^+$ are directly proportional to $\dot{N}_A^+$ and thus, directly proportional to $\dot{V}_{inert}$. Using equation (A.9) to rewrite equations (A.3) and (A.8), two expressions for τ and c_A^+ depending on $\dot{V}_{inert}$ and the real factor *RF* can be written.

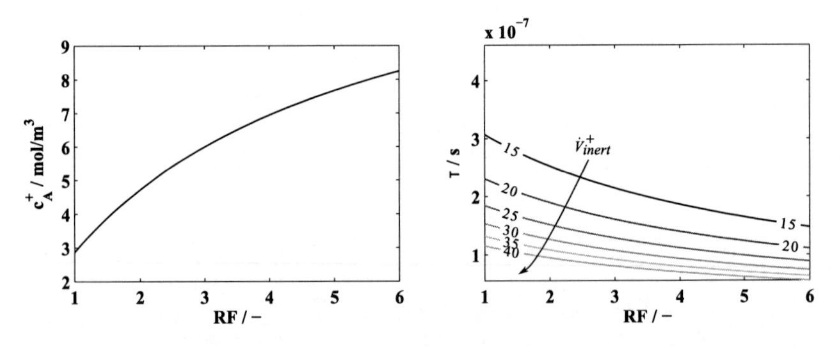

Figure A.2: Variation of c_A^+ (left) and of τ (right) against *RF* for several inert volume flows $\dot{V}_{inert}^+$ in Nl/min.

As it can be seen in figure A.2, the inert volume flow term in the definition of c_A^+ cancels, being the variation of the latter only dependant on *RF*. This is not the case for τ, which decreases for an increasing volume flow as well as an increasing *RF*. According to equation

(A.7), increasing the inlet flow reduces the term a, whereas the right side of the equation remains constant. In order to fulfill the balance, the RF applied needs to be reduced, in order to sink the concentration of the fuel. This explains the behaviour observed experimentally, in which larger values of $\dot{V}_{inert}$ correlate with a reduction of the RF applied in order to mantain the maximal reheating temperature.

A.2 Reverse flow operation

Preliminary estimation of operation parameters

Table A.5: Parameter estimation for periodic operation with hydrogen as fuel at several reactor loads. *GHSV* estimated for the inlet and outlet flow conditions in relation to the volume of active packing.

Thermal power (kW_{LHV,H_2})	**H_2-Yield** (Nm^3/h)	**GHSV** (h^{-1})	w_{therm} (cm/s)	Q_{prod} (kW)	$\dot{V}_{inert,reg}$ (Nl/min)
2	0.7	1360.5 - 2040.7	0.02	0.5	15.4
4	1.3	2721.0 - 4081.4	0.05	1.0	30.7
6	2.0	4081.5 - 6122.2	0.07	1.5	46.0

Experimental runs with H_2 as fuel

Table A.6: Parameters characterizing operation represented in figure 5.8.

Thermal power	2.37	2.97	kW_{LHV,H_2}
τ_{prod}	300	200	s
τ_{reg}	300	200	s
$\dot{V}_{CH_4}$	4	5	Nl/min
$\dot{V}_{inert}$	15	20	Nl/min
RF_1	3.5	3.75	–
RF_2	2.25	2.25	–
RF_3	2.0	2.0	–

Experimental runs with PSA off-gas as fuel

Table A.7: Performance comparison for two different loads using PSA off-gas as fuel.

Thermal power	2.37	2.97	4	kW_{LHV,H_2}
τ_{prod}	300	200	100	s
τ_{reg}	300	200	200	s
$\dot{V}_{CH_4}$	4	5	6.8	Nl/min
$\dot{V}_{inert}$	10	15	15	Nl/min
RF_1	3.75	3.75	3.2	–
RF_2	2.0	2.0	2.5	–
RF_3	1.0	1.0	1.75	–
X_{exp}/X_{equil}	92.7 / 98.5	96.0 / 99.6		%
η_{therm}	49.7	48.4		%

Appendix B

Technical drawings

B.1 Elements of the combustion chamber

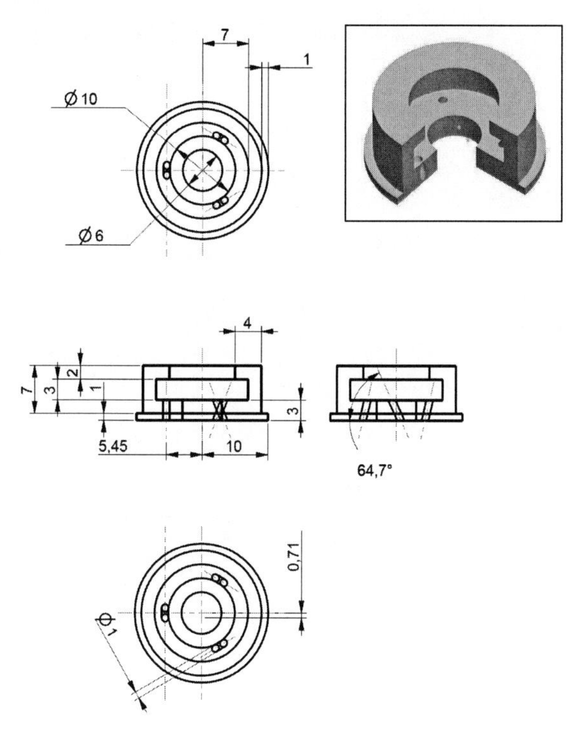

Figure B.1: Detailed technical drawing of an air nozzle as defined by Eichhorn in [76] and 3D representation thereof. Openings in the center of the top and bottom covers enable the introduction of air to the current and a further nozzle via a central duct.

B.2 Elements of the single chamber reactors

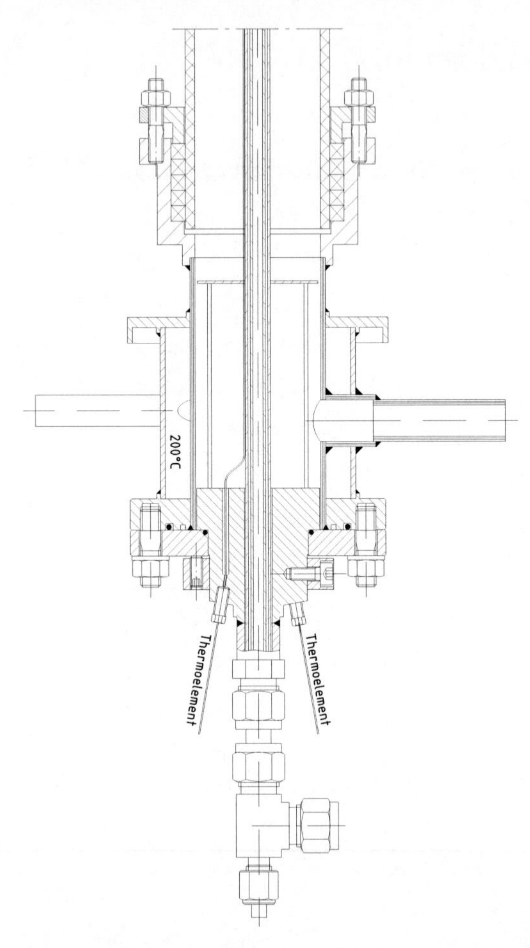

Figure B.2: Detailed technical drawing of the reactor head used to seal the end of the quartz glass reactor tube [62].

B.3 Reverse-flow reformer

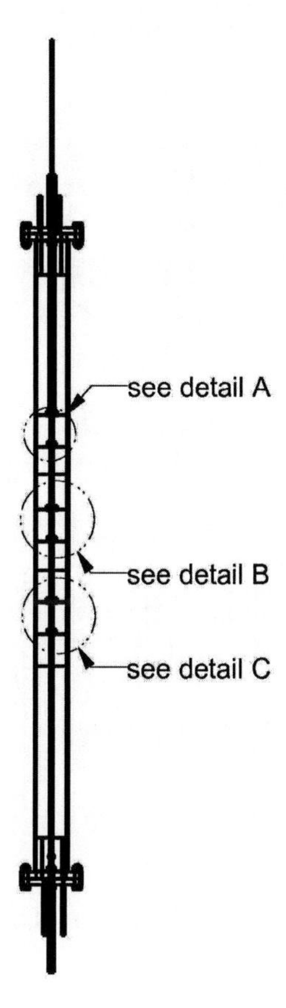

Figure B.3: Detailed technical drawing of the RFR.

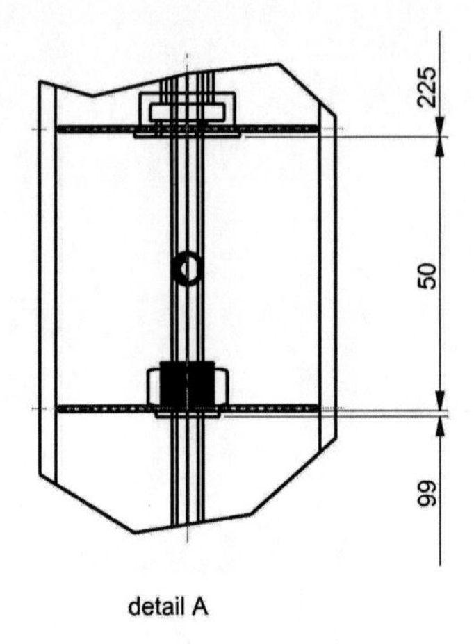

Figure B.4: Detailed technical drawing of detail A (s. figure B.3).

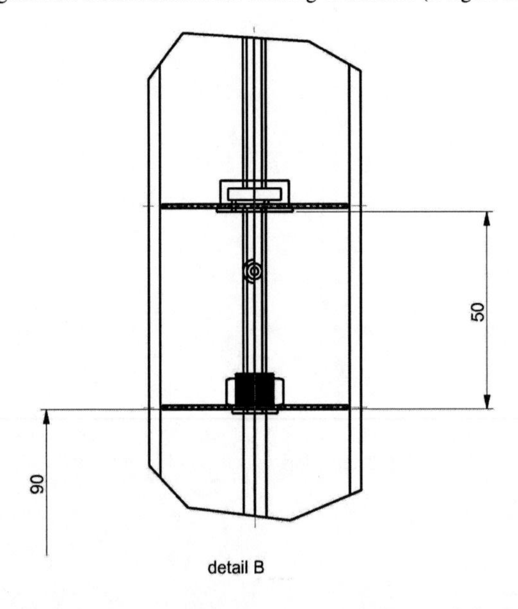

Figure B.5: Detailed technical drawing of detail B (s. figure B.3).

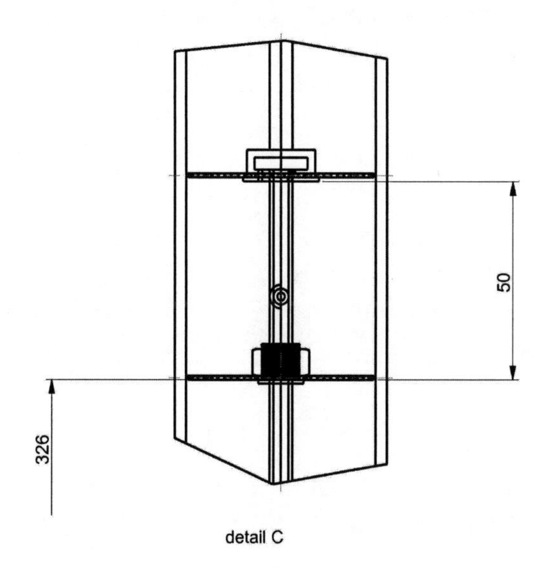

Figure B.6: Detailed technical drawing of detail C (s. figure B.3).

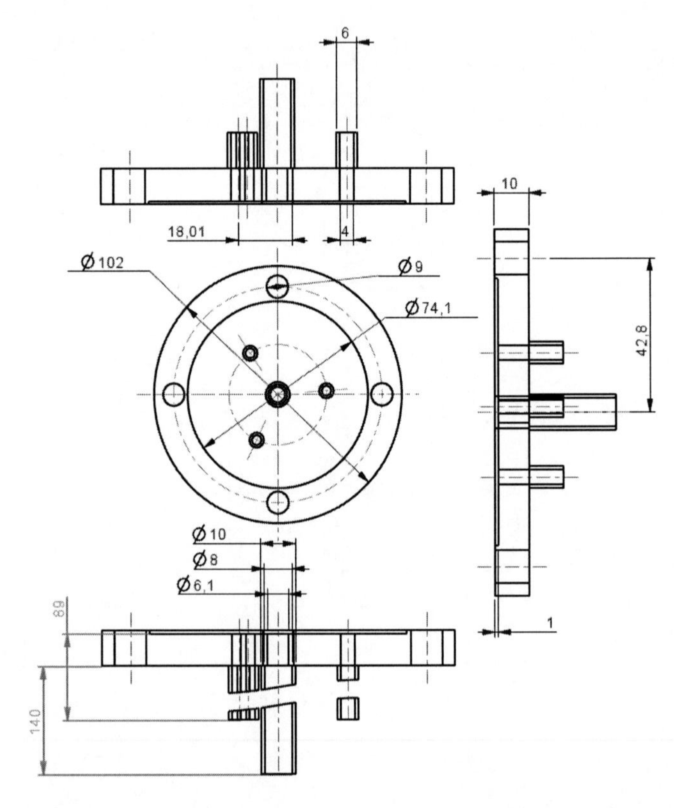

Figure B.7: Detailed technical drawing of the reactor end based on flanges.

Appendix C

Model parameters and physical properties

C.1 Model and simulation parameters set

C.1.1 Geometric dependant parameters

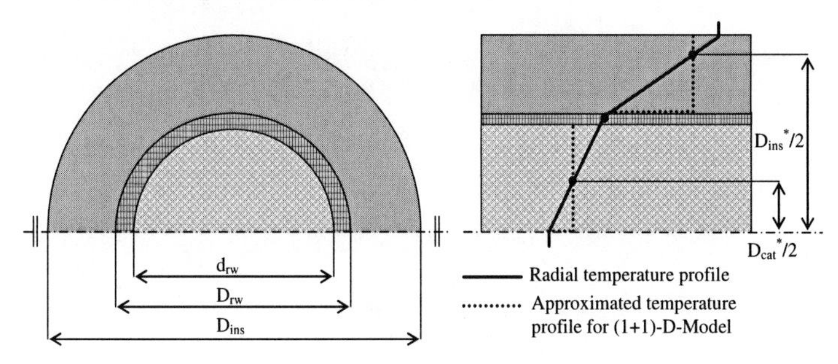

Figure C.1: Schematic representation of the system geometry and the approximated radial temperature profile.

Volume fractions of each phase in the system considered as well as the average geometric surfaces to estimate heat exchange between the different phases are dependant on the reactor geometry. The equations used based on the nomenclature used in figure C.1 are next listed.

$$A_{total} = \pi \cdot \left(\frac{D_{ins}}{2}\right)^2 \tag{C.1}$$

$$\Phi_{rea} = \frac{\pi \cdot \left(\frac{d_{rw}}{2}\right)^2}{A_{total}} \tag{C.2}$$

$$\Phi_{rw} = \frac{\pi \cdot \left[\left(\frac{D_{rw}}{2}\right)^2 - \left(\frac{d_{rw}}{2}\right)^2\right]}{A_{total}} \tag{C.3}$$

$$\Phi_{ins} = \frac{\pi \cdot \left[\left(\frac{D_{ins}}{2}\right)^2 - \left(\frac{D_{rw}}{2}\right)^2\right]}{A_{total}} \tag{C.4}$$

In turn, the specific heat transfer areas between phases $a^v_{i,j}$ read as follows

$$a^v_{gas,rw} = \frac{\pi \cdot d_{rw} \cdot L}{V_{reactor}} \tag{C.5}$$

$$a^v_{rw,ins} = \frac{\pi \cdot D_{rw} \cdot L}{V_{reactor}} \tag{C.6}$$

$$a^v_{ins,amb} = \frac{\pi \cdot D_{ins} \cdot L}{V_{reactor}} \tag{C.7}$$

with

$$V_{reactor} = \pi \cdot \left(\frac{D_{iso}}{2}\right)^2 \cdot L \tag{C.8}$$

Additionally, for the estimation of the heat transfer coefficients between phases (sec. C.1.3), the estimation of the geometrically average diameter of the different phases is required

$$D^*_{cat} = \sqrt{\frac{d^2_{rw}}{2}} \tag{C.9}$$

$$D^*_{ins} = \sqrt{\frac{D^2_{iso} + D^2_{rw}}{2}} \tag{C.10}$$

C.1.2 Physical and thermodynamic data of the bulk phases

Solid phases

The parameters of the solid phase summarized in table C.1 are taken as constant. The values for the packing have been chosen according to the data provided for aluminium oxide in [123].

Table C.1: Physical data of the solid phases involved in the system (cf. fig. 6.1).

Phase	$(\rho \cdot c_p)_j$ $\left(\frac{kJ}{m^3 \cdot K}\right)$	λ_j $\left(\frac{kW}{m \cdot K}\right)$	$\lambda_{radial,j}$ $\left(\frac{kW}{m \cdot K}\right)$
Insulation	80.0	1.0E-04	-
Reactor wall	4.71E+03	30.0E-03	-
Packed bed	2.92E+03	1.30E-02	7.0E-4

Gas phase

Gas properties are assumed to be state dependant. Density ρ_{gas} is calculated using the ideal gas law as equation of state. This is justified considering the high operation temperatures occurring during operation of the RFR.

Heat capacities $c_{p,j}$ of pure components are modelled based on the NASA-polynomials [124]. Each component is provided with two polynomial coefficient sets, describing the temperature range below and above 1000 K respectively. In order to avoid discontinuities, new coefficients have been fitted to describe a single temperature span, ranging from 200 to 1300 K (cf. tab. C.2). The resulting polynom is of the form

$$cp_j(T) = a_0 + a_1 \cdot z(T) + a_2 \cdot z(T)^2 + a_3 \cdot z(T)^3 + a_4 \cdot z(T)^4 + a5 \cdot z(T)^5 \qquad \text{(C.11)}$$

$$z(T) = \frac{T - 800}{293.0027}\ (K)$$

Table C.2: Adjusted coefficients for equation (C.11). $c_{p,j}$ in (J/kg · K).

	CH_4	H_2O	H_2	CO	CO_2	O_2	N_2
a_0	3990,7	2149,9	14680,4	1138,3	1168,8	1054,7	1120,9
a_1	942,2	202,9	412,8	76,6	112,3	62,9	76,1
a_2	-132,9	8,4	200,0	-6,9	-29,9	-19,9	1,4
a_3	19,2	-0,8	-138,8	-9,3	6,7	2,2	-14,3
a_4	17,3	1,0	-33,9	2,3	-1,2	2,9	0,4
a_5	-12,5	-1,7	44,1	0,9	0,1	-1,1	2,2

The evaluation of gas viscosity and conductivity is also required for the estimation of transport parameters between phases. Due to the steep temperature and concentration gradients occurring in the system during operation, the physical properties and thus the transport parameters may exhibit significant changes that difficult the numerical solution of the model. Therefore, both viscosity and conductivity are averaged for the temperature range of operation in the process (200 – 1 000 °C). The significant composition differences between both temperatures and between production and regeneration periods are also accounted for.

The viscosity of the pure substances is estimated based on the method proposed in [125], whereas the viscosity of the gas mixture follows the method proposed by Wilke for low pressure gases and the approximation from Herning and Zipperer for the binary parameters [108].

The conductivity of pure components is estimated based on the modified Eucken correlation, whereas the gas mixture property is based on the expression proposed by Wassiljewa after modification of Mason and Saxena [108].

C.1.3 Transport properties

Axial transport

The dispersive terms in the mass and energy gas balances accounting for deviations from the plug-flow regime are calculated based on the expression for the Peclet number (*Pe*). The later is defined as

$$\lambda_{eff} = \frac{\dot{m} \cdot c_{p,gas} \cdot d_p}{Pe} \tag{C.12}$$

and

$$D_{eff} = \frac{\dot{m} \cdot d_p}{Pe} \tag{C.13}$$

for the axial thermal and molecular diffusivities respectively. For design purposes, a *Pe* value of 2 can be assumed [126].

Heat transfer coefficients

The consideration of state and hence, position dependant heat transfer coefficients between the gas phase and the packing (α_{cat}) and the reactor wall ($\alpha_{gas,rw}$), strongly hampers the solvability of the mathematical model. In order to overcome this issue, constant values for production and regeneration are considered respectively.

The heat transfer coefficient between the packed bed and the gas phases α_{cat} is estimated using the approach proposed in [127] for the heat transfer coefficient between particles and fluids in packings. The estimation of average Prandtl (*Pr*) and Reynolds (*Re*) numbers contained in the Nusselt (*Nu*) correlation are estimated independently for the production and regeneration steps. In both cases, an average temperature of 873.15 K has been taken as reference. In order to account for the effect of the mass flow density, following functions of α_{cat}, valid in the range $0.5\,\mathrm{kg/m^2/s} \leq \dot{m} \leq 3.0\,\mathrm{kg/m^2/s}$, have been adjusted and implemented

$$\alpha_{cat,prod} = 1.229 \cdot \dot{m}^{0.4987} \frac{kW}{m^2 \cdot K} \tag{C.14}$$

$$\alpha_{cat,reg} = 0.393 \cdot \dot{m}^{0.4954} \frac{kW}{m^2 \cdot K} \tag{C.15}$$

Analogously, for the heat transfer coefficient between gas and the reactor wall, the correlation proposed by Dixon et al. is used [128]. The latter is based on mass transfer experiments and application a heat and mass transfer analogy. The adjusted and implemented functions read as follows

$$\alpha^{prod}_{gas,rw} = 0.853 \cdot \dot{m}^{0.8020} \frac{kW}{m^2 \cdot K} \tag{C.16}$$

$$\alpha^{reg}_{gas,rw} = 0.280 \cdot \dot{m}^{0.7909} \frac{kW}{m^2 \cdot K} \tag{C.17}$$

Additionally, heat transfer in the radial direction between the reactor wall and an insulation layer is considered. The transport coefficients are geometry dependant (see sec. C.1.1 and fig. C.1) and can be defined as follows

$$k^{gas,rw} = \frac{\alpha_{gas,rw}}{1 + \frac{\alpha_{gas,rw} \cdot d_{rw}}{2 \cdot \lambda_{radial,cat}} \cdot ln \frac{d_{rw}}{D^*_{cat}}} \tag{C.18}$$

$$k^{rw,ins} = \frac{1}{D_{rw}} \cdot \frac{2 \cdot \lambda_{ins}}{ln \cdot \frac{D^*_{ins}}{D_{rw}}} \tag{C.19}$$

$$k^{ins,amb} = \alpha_{ins,amb} \tag{C.20}$$

The convective heat transfer coefficient to the environment is assumed to be constant with a value of $1.5 \times 10^{-2}\,\mathrm{kW/m^2 \cdot K}$.

C.1.4 Simulation parameters set

As already mentioned, the geometric parameters of the reactor model bear on the characteristics of the reactor used in the experimental setup and described in section 5.1. The simulation parameters differ however from the experimental ones, since the former ones are chosen in order to operate in a range where heat and mass transport occur mainly due to flow mixing. For packed beds of spherical packing, this is assumed to be the case at superficial velocities corresponding to Peclet numbers above the critical value of $Pe_{crit} = 100$ [127].

The minimal mass flow density has been set to $1.0\,\mathrm{kg/m^2 \cdot s}$, which delivers *Pe* values close to the critical region.

C.1.5 Derivation of the combustion chamber control strategy

The dependencies within the several flows are next detailed according to the notation used in figure 6.2 and the stationarity condition in an adiabatic combustion chamber described as

$$0 = \dot{M}^{+} \cdot c_p^{+} \cdot T_{gas}^{+} - \dot{M}_{out} \cdot c_{p,out} \cdot T_{gas,out} + q_{combustion} \tag{C.21}$$

For an i-indexed chamber, the inlet term is defined as

$$\sum_{m=1}^{M} \dot{M}^{+} = \dot{M}_{in,i-1} + \dot{M}_{fuel,i} + \dot{M}_{air,i} \tag{C.22}$$

for the mass balance, and as

$$\begin{aligned} \dot{M}^{+} \cdot c_p^{+} \cdot T_{gas}^{+} = {} & \dot{M}_{in,i-1} \cdot c_{p,in} \cdot T_{gas,i-1} \\ & + \dot{M}_{fuel,i} \cdot c_{p,fuel} \cdot T_{fuel,i} \\ & + \dot{M}_{air,i} \cdot c_{p,air} \cdot T_{air,i} \end{aligned} \tag{C.23}$$

for the energy balance.

The inlet terms can be further detailed. The term $\dot{M}_{in,i-1}$ contains variable amounts of H_2O, CO_2 and N_2 depending on the main stream fed to the reactor during regeneration as well as on the operation conditions in the combustion chambers located before chamber i. Thus, the inert transport term in the energy balance (C.23) can be written as

$$\begin{aligned} \dot{M}_{in,i-1} \cdot c_{p,in} \cdot T_{gas,i-1} = {} & \dot{M}_{N_2,i-1} \cdot c_{p,N_2,i-1} \cdot T_{gas,i-1} \\ & + \dot{M}_{H_2O,i-1} \cdot c_{p,H_2O,i-1} \cdot T_{gas,i-1} \\ & + \dot{M}_{CO_2,i-1} \cdot c_{p,CO_2,i-1} \cdot T_{gas,i-1} \end{aligned} \tag{C.24}$$

The fuel consists of a given mixture of the components $j = CH_4$, CO, CO_2, H_2 and N_2. Thus, the transport term in equation (C.23) can be rewritten as

$$\dot{M}_{fuel,i} \cdot c_{p,fuel} \cdot T_{fuel,i} = \dot{M}_{fuel,i} \cdot \left(\sum_{j=1}^{5} w_{fuel,j} \cdot c_{p,j} \right) \cdot T_{fuel,i} \tag{C.25}$$

where $w_{fuel,j}$ defines the mass fraction of each component j in the fuel. Equally, the term for the air flow in the mass balance can be rewritten as

$$\begin{aligned}\dot{M}_{air,i} &= \dot{M}_{O_2,i} + \dot{M}_{N_2,i} \\ &= N_{O_2,i} \cdot MW_{O_2} + (0.79/0.21) \cdot N_{O_2,i} \cdot MW_{N_2}\end{aligned} \tag{C.26}$$

and taking the stoichiometry of the combustion reactions (2.7) to (2.9) into account

$$\dot{M}_{O_2,i} = \underbrace{MW_{O_2} \cdot \left(\frac{1}{2} \cdot \frac{w_{fuel,H_2}}{MW_{H_2}} + 2 \cdot \frac{w_{fuel,CH_4}}{MW_{CH_4}} + \frac{1}{2} \frac{w_{fuel,CO}}{MW_{CO}} \right)}_{k_{O_2}} \cdot \dot{M}_{fuel,i} \tag{C.27}$$

$$\dot{M}_{N_2,i} = \underbrace{3.76 \cdot MW_{N_2} \cdot \left(\frac{1}{2} \cdot \frac{w_{fuel,H_2}}{MW_{H_2}} + 2 \cdot \frac{w_{fuel,CH_4}}{MW_{CH_4}} + \frac{1}{2} \frac{w_{fuel,CO}}{MW_{CO}} \right)}_{k_{N_2}} \cdot \dot{M}_{fuel,i} \tag{C.28}$$

Introducing equations (C.27) and (C.28) in (C.26), the air flow term in equation (C.23) can be written in terms of the fuel mass flow

$$\dot{M}_{air,i} \cdot c_{p,air} \cdot T_{air,i} = \dot{M}_{fuel,i} \cdot (k_{O_2} \cdot c_{p,N_2} + k_{N_2} \cdot c_{p,N_2}) \cdot T_{air,i} \tag{C.29}$$

The oultet streams are defined analogous to equations (C.22) and (C.23). As already mentioned in the model assumptions, combustion in each chamber occurs stoichiometrically and total conversion of the fuel is achieved. The combustion products are thus H_2O and CO_2 and together with the nitrogen being introduced with the combustion air and the inert inlet itself, the terms for the mass and energy balances are respectively formulated as follows:

$$\sum_{n=1}^{N} \dot{M}_{out} = \dot{M}_{product,i} + \dot{M}_{in,i} \tag{C.30}$$

$$\begin{aligned}\dot{M}_{out} \cdot c_{p,out} \cdot T_{gas,out} &= \dot{M}_{product,i} \cdot c_{p,product,i} \cdot T_{gas,out} \\ &+ \dot{M}_{in,i} \cdot c_{p,in} \cdot T_{gas,out}\end{aligned} \tag{C.31}$$

The product flow is defined based on the amounts of H_2O and CO_2 resulting from the combustion reactions. Thus, it can be described as a function of the fuel being oxidized and the term in equation (C.30) can be written as

$$\begin{aligned}\dot{M}_{product,i} &= \dot{M}_{H_2O,i} + \dot{M}_{CO_2,i} \\ &= N_{H_2O,i} \cdot MW_{H_2O} + N_{CO_2,i} \cdot MW_{CO_2}\end{aligned} \tag{C.32}$$

with

$$\dot{M}_{H_2O,i} = \underbrace{MW_{H_2O} \cdot \left(\frac{w_{fuel,H_2}}{MW_{H_2}} + 2 \cdot \frac{w_{fuel,CH_4}}{MW_{CH_4}} \right)}_{k_{H_2O}} \cdot \dot{M}_{fuel,i} \tag{C.33}$$

$$\dot{M}_{CO_2,i} = \underbrace{MW_{CO_2} \cdot \left(\frac{w_{fuel,CH_4}}{MW_{CH_4}} + \frac{w_{fuel,CO}}{MW_{CO}} \right)}_{k_{CO_2}} \cdot \dot{M}_{fuel,i} \tag{C.34}$$

and consequently

$$\dot{M}_{product,i} \cdot c_{p,product} \cdot T_{gas,out} = \dot{M}_{fuel,i} \cdot \left(k_{H_2O} \cdot c_{p,H_2O} + k_{CO_2} \cdot c_{p,CO_2} \right) \cdot T_{gas,out} \tag{C.35}$$

The inert flow leaving the chamber is in turn defined as the inert flow entering the chamber and additionally, the nitrogen being fed with the combustion air as well as the inert substances contained in the fuel (CO_2 and N_2). Thus, mass and energy transport terms can be defined as

$$\begin{aligned}\dot{M}_{in,i} &= \dot{M}_{inert,i-1} + \dot{M}_{fuel,i} \cdot \left(w_{fuel,N_2} + w_{fuel,CO_2} \right) + \dot{M}_{N_2,i} \\ &= \dot{M}_{inert,i-1} + \dot{M}_{fuel,i} \cdot \left(w_{fuel,N_2} + w_{fuel,CO_2} + k_{N2} \right)\end{aligned} \tag{C.36}$$

and consequently equation (C.31) can be rewritten as

$$\begin{aligned}\dot{M}_{out} \cdot c_{p,out} \cdot T_{gas,out} = {} & \dot{M}_{fuel,i} \cdot \left(k_{H_2O} \cdot c_{p,H_2O} + k_{CO_2} \cdot c_{p,CO_2} \right) \cdot T_{gas,out} \\ & + \dot{M}_{fuel,i} \cdot \left(w_{fuel,N_2} + w_{fuel,CO_2} + k_{N2} \right) \cdot T_{gas,out} \\ & + \dot{M}_{N_2,i-1} \cdot c_{p,N_2,i} \cdot T_{gas,out} \\ & + \dot{M}_{H_2O,i-1} \cdot c_{p,H_2O,i} \cdot T_{gas,out} \\ & + \dot{M}_{CO_2,i-1} \cdot c_{p,CO_2,i} \cdot T_{gas,out}\end{aligned} \tag{C.37}$$

In order to determine the amount of fuel required to increase the temperature $T_{gas,i-1}$ to the desired reheating level $T_{gas,out}$, equations (C.24), (C.25) and (C.29) must be introduced in (C.23). The resulting formulation, together with equation (C.37), must be further introduced in the stationary energy balance (C.21). After solving for the fuel mass flow, following expression is obtained

$$\dot{M}_{fuel,i} = \frac{a_1 - a_2}{b_1 + b_2 + b_3} \tag{C.38}$$

$$a_1 = \dot{M}_{in,i-1} \cdot c_{p,in} \cdot T_{gas,i-1} \tag{C.39}$$

$$a_2 = \left(\dot{M}_{N_2,i-1} \cdot c_{p,N_2,i} + \dot{M}_{H_2O,i-1} \cdot c_{p,H_2O,i} + \dot{M}_{CO_2,i-1} \cdot c_{p,CO_2,i}\right) \cdot T_{gas,out} \tag{C.40}$$

$$b_1 = (k_{CO_2} + w_{fuel,CO_2}) \cdot c_{p,CO_2,i} + k_{H_2O} \cdot c_{p,H_2O,i} + (k_{O_2} \cdot 3.76 + w_{fuel,N_2}) \cdot c_{p,N_2,i}) \cdot T_{gas,out} \tag{C.41}$$

$$b_2 = -\left(3.76 \cdot k_{O_2} \cdot c_{p,N_2,air} + k_{O_2} \cdot c_{p,O_2,air}\right) \cdot T_{air,i} \tag{C.42}$$

$$b_3 = -\left(\sum_{J}^{j=1} w_{fuel,j} \cdot c_{p,j,fuel}\right) \cdot T_{fuel,i} + \Delta h_{combustion} \tag{C.43}$$

and

$$\Delta h_{combustion} = \sum_{j=1}^{3} \left(-\Delta h_{C,j}\right) \cdot \frac{w_{fuel,j}}{MW_j} \tag{C.44}$$

Appendix D

Parametric studies based on simulation

The results reported in this chapter provide additional information to the simulation results discussed in chapter 7. With help of the mathematical model, parametric studies accounting for the effects of some of the operation parameters that have not been explicitely discussed in chapter 7 are briefly reported here.

D.1 Asymmetric operation

D.1.1 Operation parameters

Reheating fuel

The possibility to operate the RFR with different fuels has been experimentally proven and discussed in section 5.3.3. As already pointed out there, the energy density per unit mass of the PSA off-gas considered is approximately an order of magnitude smaller than that of pure hydrogen. Furthermore, for the same energy input, not only larger amounts of fuel but also slightly larger amounts of air are required in case of using PSA off-gas as fuel. This outweights the lower heat capacity of the combustion products compared to those from pure hydrogen combustion (s. fig. 7.1) and is responsible for an increase of the average outlet temperature during regeneration. This can be seen in figure D.1, top, where the stationary profiles for a constant $\dot{m}_{reg}/\dot{m}_{prod}$ ratio and use of PSA off-gas and H_2 as fuel are compared.

Additionally, the fuel and air mass flows in the three combustion chambers during regeneration are plotted against the duration of the regeneration period in the lower part of figure D.1. The reheating task at the beginning of the regeneration period is assumed by the leftmost and middle chambers, since they are positioned in the reactor zone that has been primarily cooled down during the production period. The rightmost chamber, on the contrary, assumes a reheating function towards the end of the regeneration period, after the regenerative heat exchanger cools down and the temperature of the inert flow entering the chamber decreases. This trend is in very good agreement with the experimental results obtained with hydrogen as fuel (cf. fig. 5.11).

In terms of the performance indicators introduced in chapter 6, it can be stated that the somewhat larger fuel and air mass flows required when operating the RFR with PSA off-gas as fuel for a constant $\dot{m}_{reg}/\dot{m}_{prod}$ ratio are the cause of an increase of the average oulet temperature only during the regeneration period (cf. fig. D.2, b). Hence, as shown in figure D.2, a, heat losses with the product flow increase and the energetic efficiency decreases accordingly, in comparison to operation under similar conditions with H_2 as fuel. It is worth noticing that the position of the maxima in the efficiency curves is very slightly shifted to lower $\dot{m}_{reg}/\dot{m}_{prod}$ ratios since the *energy gap* left by the endothermic reaction is regenerated somewhat faster due to the larger mass flow in case of using PSA off-gas as fuel.

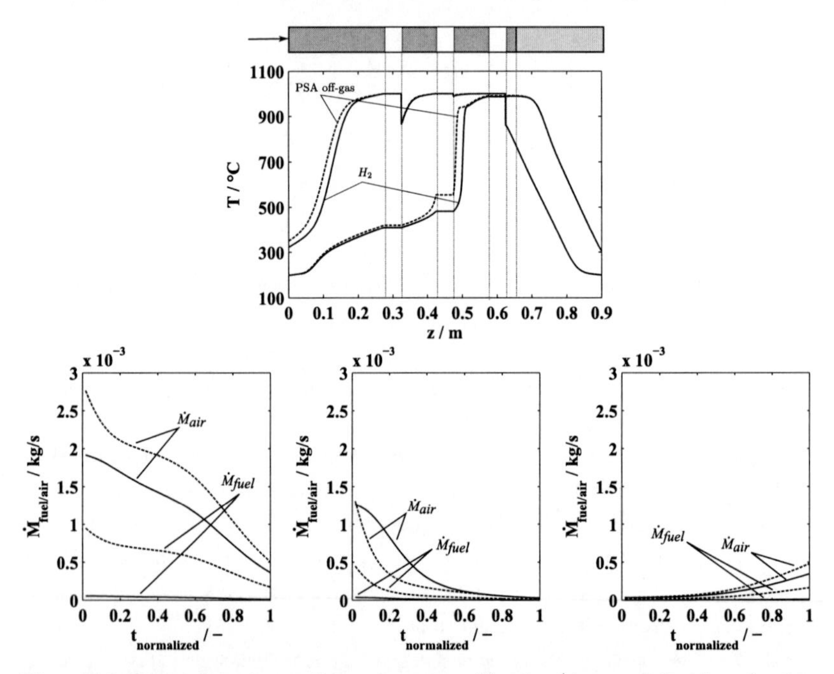

Figure D.1: Top: stationary state for $\tau_{prod} = \tau_{reg} = 60\,\mathrm{s}$, $\dot{m}_{reg}/\dot{m}_{prod} = 2.6$ and combustion of H_2 (straight lines) and PSA off-gas (dashed lines) for regeneration.
Bottom, from left to right: fuel/air mass flow rates in the left, middle and right combustion chambers during regeneration.

Based on the behaviour described and in full agreement with the experimental results discussed in chapter 5, it can be concluded that the use of H_2, or in more general terms, of fuels exhibiting large heating values and requiring low amounts of combustion air, is preferred in order to enhance the thermal efficiency of the system. This holds true whenever the heat flux during production and regeneration is balanced in the rightmost regenerative heat exchange zone of the reactor.

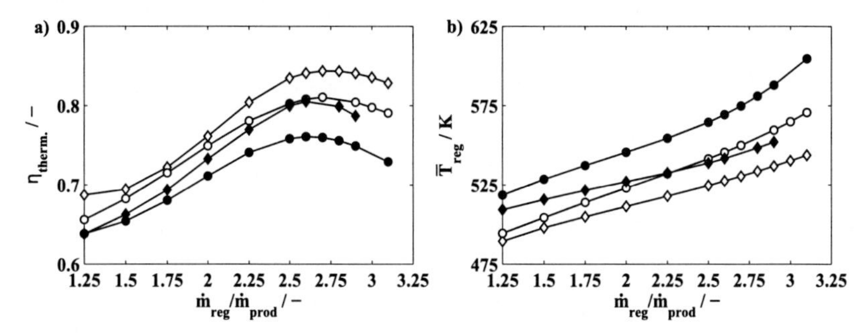

Figure D.2: a) Energetic efficiency and b) average outlet temperature during regeneration against $\dot{m}_{reg}/\dot{m}_{prod}$ for operation with PSA off-gas (black markers) and H_2 (white markers) as fuel. Cycle durations: $\tau = 60\,s$ (diamonds) and $80\,s$ (circles).

Reheating temperature

The results reported so far consider a reheating temperature T_{heat} of 1273.15 K, which is beyond the minimal temperature required to obtain full conversion of methane at atmospheric pressure ($\approx$ 1073.15 K). The effect of reducing the reheating temperature in almost 200 K can be discussed based on figure D.3, which compares the stationary temperature profiles obtained for two different $\dot{m}_{reg}/\dot{m}_{prod}$ ratios, using H_2 as fuel.

Interestingly, although higher reheating temperatures correlate with larger amounts of fuel/air, figure D.3 reveals that the temperature at the leftmost reactor boundary during regeneration is barely affected by variations in the former parameter. On the contrary, the effects become relevant only during the production period, strongly depending on the regeneration-to-production mass flow ratio. At low $\dot{m}_{reg}/\dot{m}_{prod}$ ratios (fig. D.3, a), the main flow during regeneration does not sufficiently cool down the right heat exchange zone. Thus, the effluent temperature during the subsequent production period reaches values close to T_{heat}, being the heat losses through the reactor boundaries larger in case of operating at high reheating temperatures. Conversely, at higher $\dot{m}_{reg}/\dot{m}_{prod}$ ratios (fig. D.3, b), the heat exchange zone is sufficiently cooled during the regeneration period and the reheating temperature has only an effect on the temperature slope (cf. section 2.3.2). Hence, under such operation conditions, no significant increase or reduction of the heat losses can be attained by means of simply reducing the reheating temperature. In spite thereof, provided that the distance travelled by the endothermic front increases at lower reheating temperatures, this represents an approach to enhance catalyst usage whenever the reheating temperature is such, that full conversion can be guaranteed.

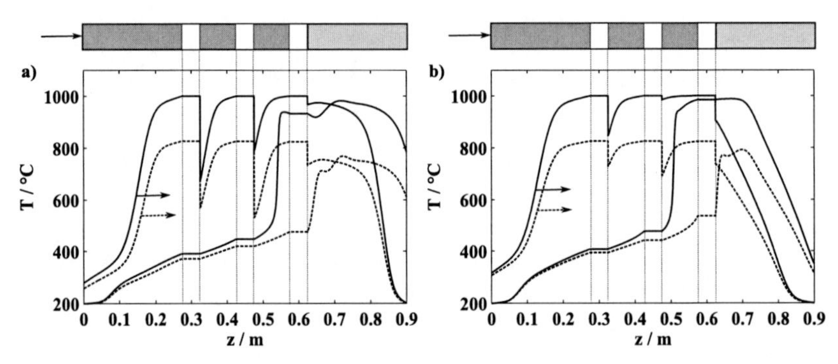

Figure D.3: Stationary temperature profiles at the beginning and end of the production period for operation at mass flow ratios $\dot{m}_{reg}/\dot{m}_{prod}$ of a) 1.75, and b) 2.5. H_2 used as fuel for reheating up to $T_{heat} = 1273.15\,K$ (straight lines) and $T_{heat} = 1100.15\,K$ (dashed lines). Switching frequency $\tau_{prod} = \tau_{reg} = 60\,s$.

In terms of the performance indicators, figure D.4, a, reveals a very slight enhancement in the thermal efficiency of the system by a reduction of T_{heat}, provided that the average temperature in the packed bed is reduced and consequently, heat losses with the reactor effluents also diminish to some extent. An improvement can be especially observed for slower switching frequencies (i.e. circles) since the weighting of the average product flow temperatures in the cumulative heat losses over time increases proportionally to the duration of the cycle. However, for faster switching frequencies this effect is only noteworthy at low mass flow ratios, whereas the difference diminishes or almost vanishes in the region of maximal efficiency (or minimal adiabatic temperature increase, fig. b).

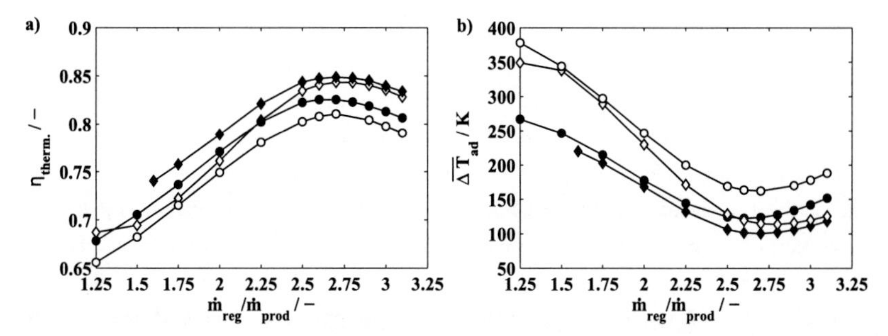

Figure D.4: RFR performance indicators against $\dot{m}_{reg}/\dot{m}_{prod}$ for two reheating temperatures T_{heat}. White and black markers for $T_{heat} = 1273.15\,\text{K}$ and $1100.15\,\text{K}$ respectively. Period durations of 80 s (circles) and 60 s (diamonds).
a) Energetic efficiency.
b) Average adiabatic temperature increase of the product flow.

Heat losses

As already observed during operation of the RFR experimental setup, heat losses are especially noteworthy in the combustion chambers. Apart from radiation losses, which are difficult to account for, the strong turbulence responsible for good mixing enhances convective heat transfer with the reactor wall and hence, with the environment. In order to emulate heat losses in the combustion chamber, the fuel heating value can be reduced while the combustion reaction stoichiometry remains unchanged. With such a simplification, heat losses are accompanied by an increase in the fuel and air mass flows required to reach a given reheating temperature in the system. The effect is analogous to that discussed for variations in the reheating fuel based on figure D.1. In this case, conversely to the case of using PSA off-gas instead of H_2 as fuel, a larger mass flow during regeneration implies a proportional increase of the heat flux, since no compensating effect through variations in the heat capacity of the combustion effluents can be expected.

D.1.2 Structure of the packed bed

Active bed length

The length of the fixed bed has only an influence on the performance of the RFR if it is operated in a regime, at which the reactants conversion is sensitive to changes in the residence time. This case is discussed by Kulkarni et al. in [37]. The mathematical model used in the study referenced considers mass transfer limitations between the gas and solid bulk phases and within the catalyst pores [60]. Thus, conversion and consequently efficiency, become directly proportional to the residence time and hence, to the reactor length.

In the current study, however, mass transfer resistances are neglected. Furthermore, spontaneous, fully conversion of the reactants is considered during the regeneration period. Likewise, kinetics considered during the production semicycle result in large rates and thermodynamic equilibrium is reached at every temperature. Thus, conversion is decoupled from residence time and variations in the reactor length do not directly affect the reactor efficiency, whenever the mass flows or the period durations are proportionally adjusted in order to maintain the ratio between production and regeneration cycles unchanged.

Heat exchanger length

Although both reactor ends are intended to act as regenerative heat exchangers, only the rightmost inert section is considered at this stage. The effect of deviations in its length are discussed based on figure D.5. The performance of a reactor with an inert section length of $L_{inert} = 275\,\text{mm}$ is taken as reference and is compared to two different configurations, accounting for a longer and a shorter inert section respectively.

The efficiency curves in figure D.5, c, and in particular that of the reference case (circles), prove the adequacy of the short cut method discussed in section 2.3.2 for an accurate design of the heat exchanger section. Accordingly, overdimensioning its length does not provide any significant improvement in the energy recovered (cf. curves with circles and diamods as markers). On the contrary, reducing the inert section length to its half has a considerable negative effect (cf. curve with squares as markers). The reason therefor are the larger heat losses that play a role at large $\dot{m}_{reg}/\dot{m}_{prod}$ ratios (see ΔT_{out} in fig. b), provided that at low mass flow ratios the inert section is generally underdimensioned independently of its length (s. fig. D.5 a). It is also relevant mentioning that variations in the rightmost heat exchanger length have none or only a very slight effect in the temperature profiles established in the catalytic reactor zones under periodic steady state (cf. figure D.5, a). Only for large $\dot{m}_{reg}/\dot{m}_{prod}$ ratios and underdimensioned heat exchange sections, a significant variation of the inlet flow temperature in the rightmost combustion chamber during regeneration (ΔT_{in} in figure b) has an effect on the fuel/air mixture fed into the system. This can be observed in the very slight variation of the temperature profiles and particularly in the position of the endothermic reaction front.

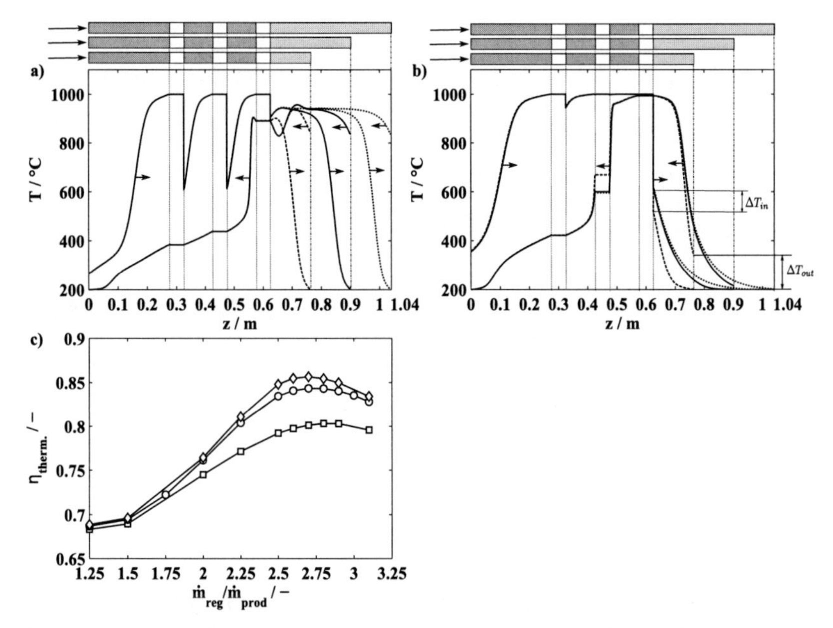

Figure D.5: Comparison of operation with varying inert heat exchange section length. Operation parameters: $T_{heat} = 1273.15\,\text{K}$, H_2 as fuel and $\tau_{prod} = \tau_{reg} = 60\,\text{s}$.
Top: Temperature profiles at the beginning and end of the production period for $L_{inert} = 412.5\,\text{mm}$ (dotted lines), $L_{inert} = 275\,\text{mm}$ (straight lines) and $L_{inert} = 137.5\,\text{mm}$ (dashed lines). Figure a) and b) for $\dot{m}_{reg}/\dot{m}_{prod} = 1.5$ and $\dot{m}_{reg}/\dot{m}_{prod} = 3.1$ respectively.
Bottom: Efficiency against $\dot{m}_{reg}/\dot{m}_{prod}$ ratio for $L_{inert} = 412.5\,\text{mm}$ (diamonds), $L_{inert} = 275\,\text{mm}$ (circles) and $L_{inert} = 137.5\,\text{mm}$ (squares).

D.2 Symmetric operation

Production and regeneration duration

The ratio between the production and regeneration time span as well as the duration of a complete cycle are parameters that can be directly variated in order to increase the temperature level in the system and thus, its productivity. Figure D.6 represents an extended version of figure 7.15 and depicts the effects of such variations in the operation parameters.

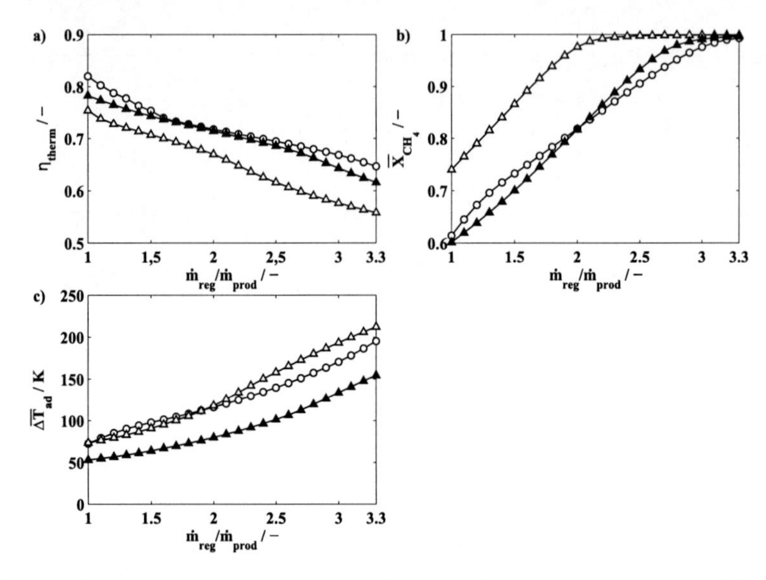

Figure D.6: Comparison of performance indicators against $\dot{m}_{reg}/\dot{m}_{prod}$ for three different operation configurations regarding τ_{reg}/τ_{prod} ratios. a), b) and c) represent thermal efficiency, average conversion and adiabatic temperature increase of the reactor effluent, respectively.
White circles: $\tau_{prod} = 20\,\text{s}$; $\tau_{reg}/\tau_{prod} = 2$.
White triangles: $\tau_{prod} = 20\,\text{s}$; $\tau_{reg}/\tau_{prod} = 3$.
Black triangles: $\tau_{prod} = 15\,\text{s}$; $\tau_{reg}/\tau_{prod} = 3$.

An increase in the regeneration to production duration ratio (τ_{reg}/τ_{prod}) causes an almost parallel reduction of the efficiency curve over $\dot{m}_{reg}/\dot{m}_{prod}$. This is observed when comparing the thermal efficiency curves in figure D.6, a, for operation with $\tau_{reg}/\tau_{prod} = 2$ (white circles) and $\tau_{reg}/\tau_{prod} = 3$ (white triangles). The explanation therefor can be found in the longer time span during which hot effluents leave the reactor, even though the adiabatic temperature increase of the latter is not significantly larger for increased τ_{reg}/τ_{prod} ratios (cf. figure c). Simultaneously, for a given $\dot{m}_{reg}/\dot{m}_{prod}$ ratio, the average conversion achieved increases if larger regeneration periods are applied (figure b), since larger energy amounts are released in the combustion chamber and convectively transported. Hence, the temperature level in the transition zone between catalytic and inert packing is substantially higher and thus, the reactor productivity larger. Even though the shift of the efficiency and conversion curves is not exactly parallel to the reference curve (circles), it could be stated that a reduction of the $\dot{m}_{reg}/\dot{m}_{prod}$ can be effectively compensated with an increase in the τ_{reg}/τ_{prod} ratio, without any noteworthy detrimental in the reactor performance.

Figure D.6 further enables analyzing the effect that a reduction of the cycle duration has on the reactor performance. The curves with triangles as markers represent both operation regimes with $\tau_{reg}/\tau_{prod} = 3$. However, whereas the curve with white markers corresponds to production steps of $\tau_{prod} = 20\,\text{s}$, the curve with black ones corresponds to $\tau_{prod} = 15\,\text{s}$. Hence, the duration of a complete cycle is reduced in 25%. An increase in the switching frequency goes accompanied by an increase in the thermal efficiency and a reduction in productivity (s. figures D.6, a and b, respectively). Moreover, in the specific case under analysis, the reactor performance gets close to that achieved with longer cycles and smaller τ_{reg}/τ_{prod} ratios (curves with circles as markers). At the same time, the adiabatic temperature increase of the reactor effluents is significately reduced, so that operability of the reactor is accordingly improved.

Figure D.7 shows the temperature profiles at the beginning and at the end of the production and regeneration steps in one direction. Representations on the top correspond to operation with shorter cycles ($\tau_{prod} = 15\,\text{s}$ and $\tau_{reg}/\tau_{prod} = 3$), whereas those on the bottom correspond to longer cycles ($\tau_{prod} = 20\,\text{s}$) and a constant τ_{reg}/τ_{prod} ratio of 3. The plots on the left (a and c) correspond to operation with $\dot{m}_{reg}/\dot{m}_{prod} = 1.5$, whereas the ones on the right (b and d) to operation with mass flow ratios of 3. The performance of the operation regimes represented in the upper and lower representations can be respectively extracted from the curves with black and white triangles as markers in figure D.6.

Although fast switching frequencies are energetically attractive (see also section 7.1.2), catalyst usage represents generally an inconvenient. Furthermore, in the current case, the slight thermal efficiency increase obtained with a fast switching strategy at low mass flow ratios does not compensate at all the productivity decrease resulting therefrom. Increasing the switching frequency only seems to be an alternative at large $\dot{m}_{reg}/\dot{m}_{prod}$ ratios, since operability is enhanced by reducing the temperature of the reactor effluents while mantaining large conversions (see figures D.7, b and d). However, a closer analysis of the performance indicators (fig. D.6) points out, that there is no significant performance enhancement if an increase in the cycle duration is accompanied by a reduction of the mass flow ratios. As an example to this behaviour should be noticed, that the performance by operation with longer cycles (white triangles) and $\dot{m}_{reg}/\dot{m}_{prod} = 2.2$ is analogous to that with a faster switching frequency (black triangles) and $\dot{m}_{reg}/\dot{m}_{prod} = 3$. The preferred operation mode accounts for larger cycles, since the number of switching events can be accordingly reduced.

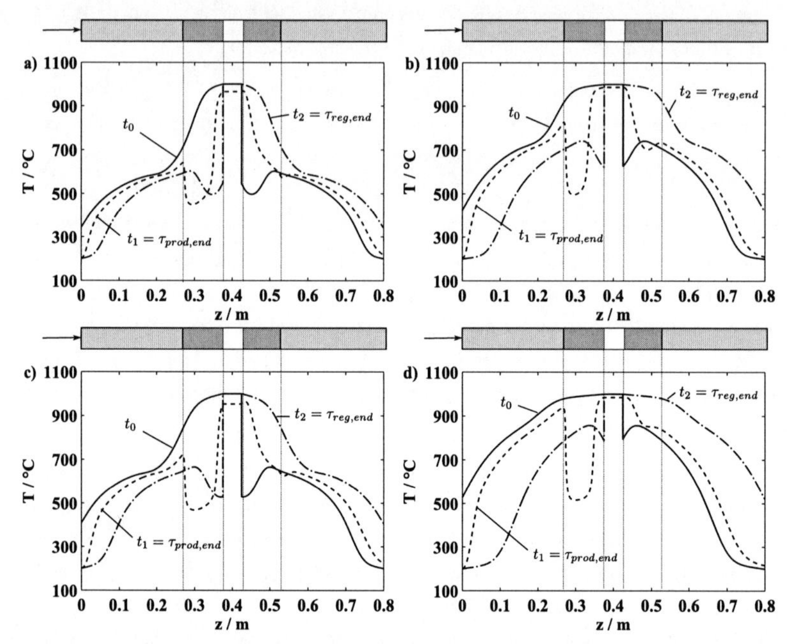

Figure D.7: Temperature profiles at the beginning (straight line) and at the end (dashed line) of the production step, and at the end of the regeneration step (dash-pointed line) in one flow direction. $\tau_{reg}/\tau_{prod} = 3$.
a) $\dot{m}_{reg}/\dot{m}_{prod} = 1.5$ and $\tau_{prod} = 15\,\text{s}$.
b) $\dot{m}_{reg}/\dot{m}_{prod} = 3.0$ and $\tau_{prod} = 15\,\text{s}$.
c) $\dot{m}_{reg}/\dot{m}_{prod} = 1.5$ and $\tau_{prod} = 20\,\text{s}$.
d) $\dot{m}_{reg}/\dot{m}_{prod} = 3.0$ and $\tau_{prod} = 20\,\text{s}$.

Appendix E

Integration of the RFR in a decentralized hydrogen production facility

Preliminary considerations to evaluate the possibility to run a decentralized hydrogen facility in an autarkic manner, using a RFR as reaction unit, are next summarized. The core aspect of this analysis is the fact that heat transfer limitations in the RFR are significantly reduced in comparison to recuperative MSR systems. Thus, the energy surpluss needed to provide the system with energy to run the endothermic reaction can be minimized. First estimations based on stationary simulations of a complete hydrogen production facility using the software AspenPlus suggest that the heating value of the residual gases leaving a PSA unit can cover the heat demand of the whole production unit.

E.1 Process description

The process flow diagram used for the present energy integration study is shown in figure E.1. Main design specification of the process is a high purity hydrogen production of 200 Nm^3/h. Accordingly, natural gas -simulated as methane- and water are fed into the process with a steam-to-carbon ratio of 3:1. Methane is available at pressures above 10-15 bar, so no compression of the flow is required. Production of 15 bar steam is assumed. Both reactants are then preheated (HE-01 and HE-02) up to 200 °C before entering the reverse-flow reactor (R-01). The latter is simulated as an equilibrium reactor operating at a fix temperature around 900 °C. The product stream is in thermodynamic equilibrium. The stream leaving the reformer at low temperatures (around 200 °C) can be redirected into a shift reactor (low temperature) without previous conditioning of the flow, in order to minimize the amount of carbon monoxide present in the mixture. The compositions of the flows leaving both the reformer and the WGS (water-gas-shift) reactor are shown in table E.1.

The flow leaving the WGS is cooled down (HE-03) before being fed to the hydrogen purification unit (PSA). Pure hydrogen is recovered at high pressure -around 15 bar- whereas the PSA off-gas can be further used as fuel. The hydrogen recovery capacity of the PSA unit determines the composition and the amount of PSA off-gas available in the process.

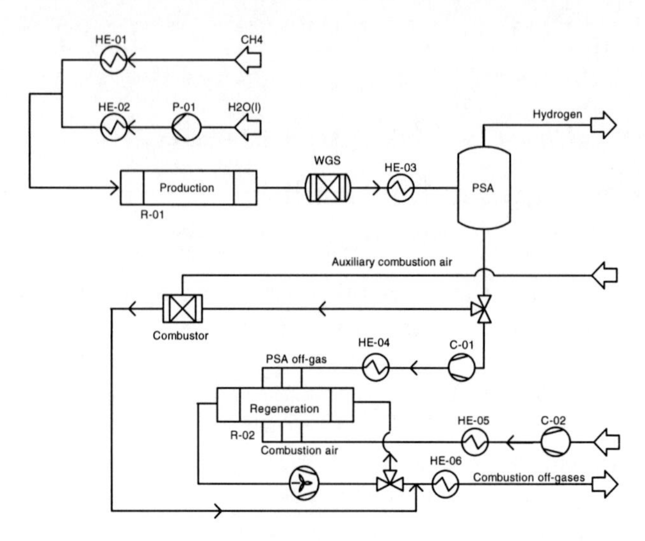

Figure E.1: Process flow diagram of a decentral H_2-production facility.

As the hydrogen production is set, its recovery in the PSA unit becomes the only degree of freedom, conditioning the remaining mass and energy flows in the whole process.

Table E.1: Composition of the product streams at the outlet of the reformer (operation temperature 900 °C) and the WGS reactor.

	Molar percentage $\left[\frac{mol_i}{mol_{tot}}\right]$				
	CH_4	H_2O	H_2	CO	CO_2
Composition after Reformer R-01	0.9	29.9	53.2	11.2	4.9
Composition after WGS-Reactor	0.9	20.1	62.9	1.4	14.7

Part of the PSA off-gases need to be compressed (C-01) in order to be used as regeneration fuel in a reverse-flow reactor operating in regeneration modus. The amount of PSA off-gas that needs to be fed to the regenerating packed bed is defined by the amount of energy that has been consumed during the reforming reaction. In order to overcome heat transfer limitations or eventual losses, a surplus in the energy demand of around 10% will be assumed. Thus, the heating value of the PSA off-gases used as fuel for the regeneration of the packed bed is specified to be 110% of the amount of energy consumed in the packed bed during a production semicycle. Combustion air is fed into the system in stochiometric relation after having been compressed (C-02).

The reverse-flow reactor operating in regenerative mode (R-02) is simulated as an adiabatic reactor where full combustion of the PSA off-gases occurs. In order to limit the maximal temperature reached during combustion, an inert flow needs to be fed to the system. The mass flow is specified in order to limit maximal temperature at 1000 °C. The combustion off-gas leaving the regenerating packed bed and containing merely carbon dioxide, water and nitrogen can be partially recirculated without prior conditioning to assume the function of the inert flow. Based on the current estimation, 34% of the combustion off-gases should be recycled, whereas the rest could be cooled down and energy contained recovered accordingly.

In case that the PSA off-gas has not been fully used for regeneration of the reverse-flow reformer, the remaining stream can be further burnt in an auxiliary combustor. The energy released can be used in the process for example, to evaporate/preheat the educts before being fed to the reformer.

The design specifications can be summarized as follows:

- The flows in the process are arranged in order to obtain a production of 200 Nm^3/h H_2.
- The whole process operates at a pressure of 10-15 bar. Therefore, consideration of pressure relief and repressurization steps between production and regeneration can be neglected.
- Chemical reactions are not kinetically limited. Thermodynamic equilibrium is reached at each reaction step.
- The reverse-flow reactor is simulated by two equilibrium steps operating stationary:
  - The first reactor R-01 reaches the equilibrium composition at the operation conditions during the production step.
  - The second reactor R-02 calculates adiabatically the heat released by the PSA off-gases combustion reaction.
- Power costs of the PSA unit are neglected, since no additional compression is required.
- Compression work is done in single step compressors followed by heat exchangers that recover the energy contained in the compressed flow.
- The auxiliary combustor is operated at atmospheric pressure as only the energy contained in the off-gases might be used. No temperature limitation is required in this case, so conventional burners can be used.

E.2 Relevant process parameters

Hydrogen yield is the main design specification of the process. The target of the study is to evaluate the potential to reduce reactants consumption and energy costs. Moreover, in an optimal configuration, the energy contained in the process streams can be internally utilized so no external energy sources are required. There are two main parameteres, which strongly influence the energy balance of the system and whose effect is summarized in the following:

1. operation conditions of R-01, and

2. hydrogen recovery, defined as the ratio between hydrogen fed to the PSA unit and that recovered as high purity product.

High hydrogen yields are favoured at high temperatures and low pressures. As first approach, only the influence of temperature in the reforming unit is taken into account. The main result of varying the reactor operation conditions is the composition of the product stream. The latter is further reacted in a WGS reactor and the leaving stream is then separated in the PSA unit. For a constant hydrogen recovery rate of 80%, the effect of increasing the reformer operation temperature can be translated in a significant reduction of the methane concentration in the PSA off-gases as shown in figure E.2.

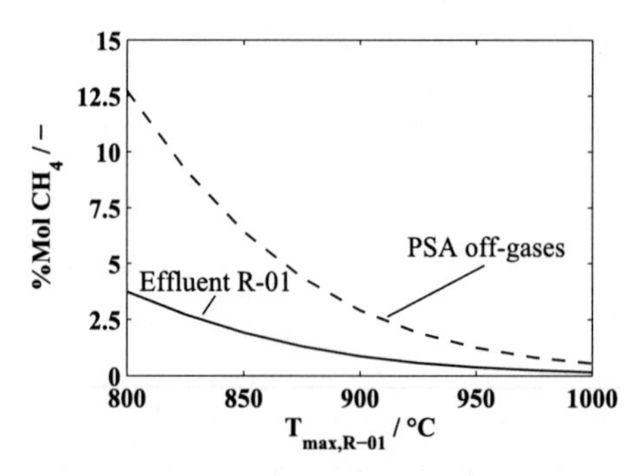

Figure E.2: Molar percentage of methane in the reformer effluent and the PSA off-gas streams as a function of the reformer operation temperature. Reforming pressure 15 bar.

The composition of the PSA off-gas directly affects the residual energy available to cover the heat demand for the reforming reaction. For a constant hydrogen production of $200\,\mathrm{Nm^3/h}$, figure E.3, left, depicts the required energy for the endothermic reaction and the available energy by full combustion of the PSA off-gases as a function of the hydrogen recovery rate in the PSA unit. The right plot, in turn, represents the relation between the energy available and that required to perform the endothermic reaction $Q_{combustion}/Q_{reaction}$. It becomes clear, that recovery rates in the range between $70-80\,\%$ are enough to sustain the RFR operation over a wide range of operation temperatures. This operation range is within realistic values for industrial PSA units, operating at adsorption and desorption pressures of 15 and 1 bar respectively [129].

Ideally, the system would be run within an operation window corresponding to

$$Q_{combustion}/Q_{reaction} \approx 1.3$$

In such a case, 30% energy surpluss could compensate heat losses, heat transfer limitations and provide other units in the process with heat. It is worth noticing that, besides the endothermic reaction itself, main heat consuming operations are natural gas preheating and steam generation and preheating.

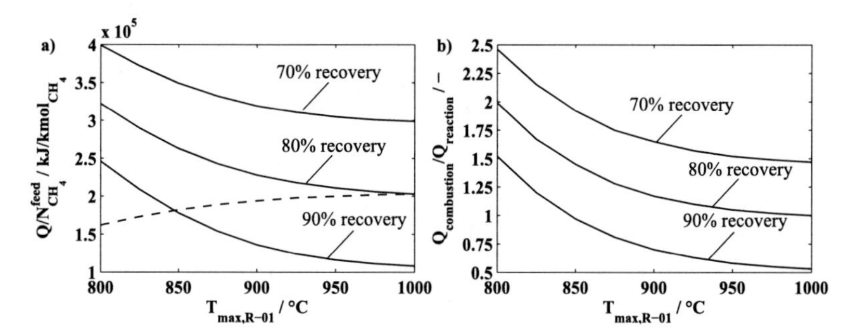

Figure E.3: Comparison of the energy consumed by the endothermic reforming reaction and the energy available by combustion of PSA off-gases as a function of the reformer operating temperature and severa hydrogen recoveries for the production of 200 Nm^3/h.
a) Required energy per kmol CH_4 processed (dashed line) and corresponding heating value available in the PSA off-gases (straight lines).
b) Available heat to the required thermal energy to perform the endothermic MSR.

Energy recovery

Besides the energy available in the off-gases leaving the auxiliary combustor, recovery of the energy contained in the WGS reactor effluent before entering the PSA unit may represent an important contribution to heat integration. Both heat sources play an important role, provided that the temperatures of the streams are far above the required entrance temperatures of the educts of 200 °C, i.e. above 300 °C for the WGS product stream and 1500 °C for the combustion products leaving the auxiliary combustor.

Bibliography

[1] Glöckler, B., Kolios, G., and Eigenberger, G. *Analysis of a novel reverse-flow reactor concept for autothermal methane steam reforming*. Chemical Engineering Science, 58(3-6):593 – 601 (**2003**).

[2] Glöckler, B., Tellaeche, C., Dieter, H., Eigenberger, G., and Nieken, U. *Propagating Endothermic Reaction Fronts and Reheating Concepts for Reverse-Flow Reformers*. In *19th International Symposium on Chemical Reaction Engineering (ISCRE-19),3.-6. September 2006, Potsdam* (**2006**).

[3] Glöckler, B., Kolios, G., Tellaeche, C., and Nieken, U. *A Heat-Integrated Reverse-Flow Reactor Concept for Endothermic High-Temperature Syntheses. Part I: Fundamentals - Short-Cut Theory and Experimental Verification of a Traveling Endothermic Reaction Zone*. Chem. Eng. Technol., 32(9):1339–1347 (**2009**).

[4] van Sint Annaland, M., Scholts, H., Kuipers, J., and van Swaaij, W. *A novel reverse flow reactor coupling endothermic and exothermic reactions. Part I: comparison of reactor configurations for irreversible endothermic reactions*. Chemical Engineering Science, 57(5):833–854 (**2002**).

[5] Agar, D. W. and Ruppel, W. *Multifunktionale Reaktoren für die heterogene Katalyse*. Chemie Ingenieur Technik, 60(10):731–741 (**1988**).

[6] Eigenberger, G. *Fixed-Bed Reactors*. In *Ullmann's Encyclopedia of Industrial Chemistry* (Wiley-VCH Verlag GmbH & Co. KGaA) (**2000**). ISBN 9783527306732.

[7] Agar, D. W. *Multifunctional reactors: Old preconceptions and new dimensions*. Chemical Engineering Science, 54(10):1299 – 1305 (**1999**). 1st International symposium on multifunctional reactors.

[8] Kolios, G., Frauhammer, J., and Eigenberger, G. *Autothermal fixed-bed reactor concepts*. Chemical Engineering Science, 55:5945–5967 (**2000**).

[9] Twigg, V. *Catalyst Handbook* (Manson Pub.) (**1996**). ISBN 9781874545354.

[10] Häussinger, P., Lohmüller, R., and Watson, A. M. *Hydrogen*. In *Ullmann's Encyclopedia of Industrial Chemistry* (Wiley-VCH Verlag GmbH & Co. KGaA) (**2000**).

[11] Kolios, G., Glöckler, B., Gritsch, A., Morillo, A., and Eigenberger, G. *Heat-integrated reactor concepts for hydrogen production by methane steam reforming.* Fuel Cells, 5(1):52–65 (**2005**).

[12] Kolios, G., Gritsch, A., Glöckler, B., and Eigenberger, G. *Enhancing Productivity and Thermal Efficiency of High-Temperature Endothermic Processes in Heat-Integrated Fixed-Bed Reactors.* In *Integrated Chemical Processes*, 1–43 (Wiley-VCH Verlag GmbH & Co. KGaA) (**2005**).

[13] Frauhammer, J., Eigenberger, G., Hippel, L., and Arnzt, D. *A new reactor concept for endothermic high-temperature reactions.* Chemical Engineering Science, 54:3661–3670 (**1999**).

[14] Kolios, G., Frauhammer, J., and Eigenberger, G. *A simplified procedure for the optimal design of autothermal reactors for endothermic high-temperature reactions.* Chemical Engineering Science, 56:351–357 (**2001**).

[15] Kolios, G., Gritsch, A., Glöckler, B., Sorescu, G., and Frauhammer, J. *Novel reactor concepts for thermally efficient methane steam reforming: modeling and simulation.* Ind. Eng. Chem. Res., 43:4796–4808 (**2004**).

[16] Ramaswamy, R., Ramachandran, P., and Dudukovic, M. P. *Recuperative coupling of exothermic and endothermic reactions.* Chemical Engineering Science, 61(2):459 – 472 (**2006**).

[17] Kolios, G., Frauhammer, J., and Eigenberger, G. *Efficient reactor concepts for coupling of endothermic and exothermic reactions.* Chemical Engineering Science, 57:1505–1510 (**2002**).

[18] Gritsch, A., Kolios, G., and Eigenberger, G. *Reaktorkonzepte zur autothermen Führung endothermer Hochtemperaturreaktionen.* Chemie Ingenieur Technik, 76(6):722–725 (**2004**).

[19] Gritsch, A., Kolios, G., Nieken, U., and Eigenberger, G. *Kompaktreformer für die dezentrale Wasserstoffbereitstellung aus Erdgas.* Chemie Ingenieur Technik, 79(6):821–830 (**2007**).

[20] Gritsch, A. *Wärmeintegrierte Reaktorkonzepte für katalytische Hochtemperatur-Synthesen am Beispiel der dezentralen Dampfreformierung von Methan*. Ph.D. thesis, Insitut für Chemische Verfahrenstechnik, Universität Stuttgart (**2008**).

[21] Nieken, U., Kolios, G., and Eigenbergr, G. *Control of the ignited steady state in autothermal fixed-bed reactors for catalytic combustion*. Chemical Engineering Science, 49(24B):5507–5518 (**1994**).

[22] Nieken, U., Kolios, G., and Eigenbergr, G. *Fixed-bed reactors with periodic flow reversal: experimental results for catalytic combustion*. Catalysis Today, 20:335–350 (**1994**).

[23] Nieken, U., Kolios, G., and Eigenbergr, G. *Limiting cases and approximate solutions for fixed-bed reactors with periodic flow reversal*. AIChE Journal, 41(8):1915–1925 (**1995**).

[24] Züfle, H., Turek, T., and Hahn, T. *Betriebsverhalten eines Festbettreaktors mit periodischer Strömungsumkehr*. Chemie Ingenieur Technik, 67(8):1008–1011 (**1995**).

[25] Khinast, J., Gurumoorthy, A., and Luss, D. *Complex dynamic features of a cooled reverse-flow reactor*. AIChE Journal, 44(5):1128–1140 (**1998**).

[26] Khinast, J., Jeong, Y. O., and Luss, D. *Dependence of cooled reverse-flow reactor dynamics on reactor model*. AIChE Journal, 45(2):299–309 (**1999**).

[27] Aubé, F. and Sapoundjiev, H. *Mathematical model and numerical simulations of catalytic flow reversal reactors for industrial applications*. Computers and Chemical Engineering, 24(12):2623 – 2632 (**2000**).

[28] Bunimovich, G. A., Strots, V. O., Matros, Y. S., and Mirosh, E. A. *Reversed flow converter: Fundamentals of the design*. SAE Technical Paper 1999-01-0459 (**1999**).

[29] Blanks, R. F., Wittring, T. S., and Peterson, D. A. *Bidirectional adiabatic synthesis gas generator*. Chemical Engineering Science, 45 (8)(8):2407–2413 (**1990**).

[30] De Groote, A. M., Froment, G. F., and Kobylinski, T. *Synthesis gas production from natural gas in a fixed bed reactor with reversed flow*. Can. J. Chem. Eng., 74(5):735–742 (**1996**).

[31] Neumann, D. and Veser, G. *Catalytic partial oxidation of methane in a high-temperature reverse-flow reactor.* AIChE Journal, 51(1):210–223 (**2005**).

[32] Snyder, J. D. and Subramaniam, B. *A novel reverse flow strategy for ethylbenzene dehydrogenation in a packed-bed reactor.* Chemical Engineering Science, 49(24, Part 2):5585 – 5601 (**1994**).

[33] Matros, Y. S. and Bunimovich, G. A. *Reverse-Flow Operation in Fixed Bed Catalytic Reactors.* Catalysis Reviews, 38(1):1–68 (**1996**).

[34] Kolios, G. and Eigenberger, G. *Styrene synthesis in a reverse-flow reactor.* Chemical Engineering Science, 54:2637–2646 (**1999**).

[35] van Sint Annaland, M. and Nijssen, R. C. *A novel reverse flow reactor coupling endothermic and exothermic reactions: an experimental study.* Chemical Engineering Science, 57:4967–4985 (**2002**).

[36] van Sint Annaland, M., Scholts, H., Kuipers, J., and van Swaaij, W. *A novel reverse flow reactor coupling endothermic and exothermic reactions: Part II: Sequential reactor configuration for reversible endothermic reactions.* Chemical Engineering Science, 57(5):855–872 (**2002**).

[37] Kulkarni, M. S. and Dudukovic, M. P. *Periodic operation of asymmetric bidirectional fixed-bed reactors: energy efficiency.* Chemical Engineering Science, 52(11):1777–1788 (**1997**).

[38] Kulkarni, M. S. and Dudukovic, M. P. *Periodic Operation of Asymmetric Bidirectional Fixed-Bed Reactors with Temperature Limitations.* Industrial & Engineering Chemistry Research, 37(3):770–781 (**1998**).

[39] Glöckler, B. *Vereinfachte Analyse eines autothermen Adsorber-Reaktors zur Wasserstoffherstellung durch Methanreformierung.* Diplomarbeit, Institut für Chemische Verfahrenstechnik, Universität Stuttgart (**2001**).

[40] Glöckler, B., Gritsch, A., Morillo, A., Kolios, G., and Eigenberger, G. *Autothermal Reactor Concepts for Endothermic Fixed-Bed Reactions.* Chemical Engineering Research and Design, 82(2):148 – 159 (**2004**). ISMR3-CCRE18.

[41] Glöckler, B., Dieter, H., Eigenberger, G., and Nieken, U. *Efficient reheating of a reverse-flow reformer - An experimental study.* Chemical Engineering Science, 62(18-20):5638 – 5643 (**2007**).

[42] Wünning, J. *Flammenlose Oxdation von Brennstoff mit hochvorgewärmter Luft.* Chemie Ingenieur Technik, 63(12):1243–1245 (**1991**).

[43] Wünning, J. G. *Flammlose Oxidation von Brennstoff.* Ph.D. thesis, RWTH Aachen (**1996**).

[44] Wünning, J. A. and Wünning, J. G. *Flameless oxidation to reduce thermal no-formation.* Progress in Energy and Combustion Science, 23(1):81 – 94 (**1997**).

[45] Wünning, J. G. *Flameless combustion and its applications.* Tech. rep., WS Inc. (**2005**). Online; accessed 26-July-2012.

[46] Stitt, E. *Multifunctional Reactors? 'Up to a Point Lord Copper'.* Chemical Engineering Research and Design, 82(2):129 – 139 (**2004**). ISMR3-CCRE18.

[47] Stegmaier, M. *Personal communication* (**2010**). Email.

[48] Piña, J., Schbib, N. S., Bucalá, V., and Borio, D. O. *Influence of the Heat-Flux Profiles on the Operation of Primary Steam Reformers.* Industrial & Engineering Chemistry Research, 40(23):5215–5221 (**2001**).

[49] Nummedal, L., Røsjorde, A., Johannessen, E., and Kjelstrup, S. *Second law optimization of a tubular steam reformer.* Chemical Engineering and Processing, 44(4):429–440 (**2005**).

[50] Wilhelmsen, i., Johannessen, E., and Kjelstrup, S. *Energy efficient reactor design simplified by second law analysis.* International Journal of Hydrogen Energy, 35(24):13219–13231 (**2010**).

[51] Rostrup-Nielsen, T. *High Flux Steam Reforming.* Tech. rep., Haldor Topsøe A/S, Denmark (**2002**).

[52] Rostrup-Nielsen, J. *Steam reforming and chemical recuperation.* Catalysis Today, 145(1-2):72–75 (**2009**).

[53] Dybkjaer, I. *Synthesis gas technology*. Tech. rep., Haldor Topsøe A/S, Denmark (**2006**).

[54] Winter-Madsen, S. and Olsson, H. *Steam reforming solutions*. Tech. rep., Haldor Topsoe A/S, Denmark (**2007**).

[55] *Topsøe HTCR Compact hydrogen units* (**2009**).

[56] Schmid, H.-P. and Wünning, J. A. *FLOX® Steam Reforming for PEM Fuel Cell Systems*. Fuel Cells, 4(4):256–263 (**2004**).

[57] Schmid, H.-P. *Personal communication* (**2010**). WS Reformer visit and documentation.

[58] Tellaeche, C. *Experimental verification of a migrating endothermic reaction front in a preheated, adiabatic packed-bed*. Diplomarbeit, Institut für Chemische Verfahrenstechnik, Universität Stuttgart (**2005**).

[59] Eigenberger, G., Kolios, G., and Nieken, U. *Thermal pattern formation and process intensification in chemical reaction engineering*. Chemical Engineering Science, 62:4825–4841 (**2007**).

[60] Kulkarni, M. S. and Dudukovic, M. P. *A bidirectional fixed-bed reactor for coupling of exothermic and endothermic reactions*. AIChE Journal, 42(10):2897–2910 (**1996**).

[61] Bernnat, J. *Detaillierte Modellierung eines autothermen Reaktors mit periodischer Strömungsumkehr zur Wasserstoffgewinnung durch Methanreformierung*. Diplomarbeit, Institut für Chemische Verfahrenstechnik, Universität Stuttgart (**2004**).

[62] Dieter, H. *Detaillierte Untersuchung des Wärmeeintrags in einen autothermen Strömungsumkehr-Festbettreaktor mit verteilter Seiteneinspeisung*. Diplomarbeit, Institut für Chemische Verfahrenstechnik, Universität Stuttgart (**2007**).

[63] *WS Wärmeprozesstechnik GmbH*. `http://www.flox.com/`. Online; accessed 26-July-2012.

[64] *WS Reformer GmbH*. `http://www.wsreformer.com/`. Online; accessed 26-July-2012.

[65] Meyer, F. *Flameless combustion*. Tech. rep., BINE Information Service (ISSN 0937-8367), 07/2006 (**2006**).

[66] Meyer, F. *Glass production - energy-efficient with low emissions*. Tech. rep., BINE Information Service (ISSN 0937-8367), 05/2008 (**2008**).

[67] Neuffer, W. *Alternative Control Techniques Document - NOx Emissions from Process Heaters (EPA-453/R-93-034)*. Tech. rep., U. S. Environmental Protection Agency, U.S. ENVIRONMENTAL PROTECTION AGENCY Emission Standards Division Office of Air and Radiation Office of Air Quality Planning and Standards Research Triangle Park, North Carolina 27711 (**1993**).

[68] Flamme, M. *Low NOx combustion technologies for high temperature applications*. Energy Conversion and Management, 42:1919–1935 (**2001**).

[69] Derudi, M., Villani, A., and Rota, R. *Sustainability of mild combustion of hydrogen-containing hybrid fuels*. Proceedings of the Combustion Institute, 31(2):3393 – 3400 (**2007**).

[70] Derudi, M., Villani, A., and Rota, R. *Mild Combustion of Industrial Hydrogen-Containing Byproducts*. Industrial & Engineering Chemistry Research, 46(21):6806–6811 (**2007**).

[71] Rink, M. *Experimentelle Untersuchung einer FLOX-Brennkammer zum kontrollierten Wärmeeintrag in einen Festbettreaktor*. Studienarbeit, Institut für Chemische Verfahrenstechnik, Universität Stuttgart (**2006**).

[72] Monz, T. *Numerische Modellierung und Untersuchung eines neuartigen Konzepts zum Wärmeeintrag in einen Strömungsumkehr-Reformer*. Studienarbeit, Institut für Chemische Verfahrenstechnik, Universität Stuttgart (**2007**).

[73] Sung, C., Law, C., and Chen, J.-Y. *Augmented Reduced Mechanisms for NO Emission in Methane Oxidation*. Combustion And Flame, 125:906–919 (**2001**).

[74] Joos, F. *Technische Verbrennung* (Springer Verlag), 1st ed. (**2006**).

[75] Zlokarnik, M. *Scale-up: Modellübertragung in der Verfahrenstechnik* (Wiley-VCH Verlag GmbH & Co. KGaA) (**2005**).

[76] Eichhorn, F. *Experimentelle Untersuchung einer Brennkammer zum effizienten Wärmeeintrag in einen Strömungsumkehr-Festbettreaktor*. Studienarbeit, Institut für Chemische Verfahrenstechnik, Universität Stuttgart (**2008**).

[77] Sendilkumar, K., Kalaichelvi, P., Perumalsamy, M., Arunagiri, A., and Raja, T., eds. *Computational Fluid Dynamic Analysis of Mixing Characteristics inside a Jet Mixer for Newtonian and Non Newtonian Fluids*. WCECS 2007 (Proceedings of the World Congress on Engineering and Computer Science 2007) (**2007**).

[78] Yousefi Amiri, T. and Moghaddas, J. S. *Experimental Study of the Mixing Time in a Jet-Mixed Gas-Liquid System*. Chemical Engineering & Technology, 33(2):327–333 (**2010**).

[79] Saravanan, Sundaramoorthy, Mohankumar, and Subramanian. *Studies on Some Aspects of Jet Mixers I: Hydrodynamics*. Modern Applied Science, 4 (3):51–59 (**2010**).

[80] Baerns, M., Hofmann, H., and Renken, A. *Chemische Reaktionstechnik*, vol. Band 1 (Georg Thieme Verlag Stuttgart), 2nd ed. (**1992**). Lehrbuch der Technischen Chemie - Band 1.

[81] Müller, W. *Untersuchung von Homogenisiervorgängen in nicht-Newtonschen Flüssigkeiten mit einem neuen bildanalytischen Verfahren*. Reihe 3, 103 (VDI-Forschungsberichte) (**1985**).

[82] Zlokarnik, M. *Stirring: Theory and practice* (Wiley-VCH Verlag GmbH) (**2007**).

[83] Veser, G. and Schmidt, L. D. *Ignition and extinction in the catalytic oxidation of hydrocarbons over platinum*. AIChE J., 42(4):1077–1087 (**1996**).

[84] Jackson, G. S., Sai, R., Plaia, J. M., Boggs, C. M., and Kiger, K. T. *Influence of H2 on the response of lean premixed CH4 flames to high strained flows*. Combustion and Flame, 132(3):503–511 (**2003**).

[85] Dagaut, P. and Nicolle, A. *Experimental and detailed kinetic modeling study of hydrogen-enriched natural gas blend oxidation over extended temperature and equivalence ratio ranges*. Proceedings of the Combustion Institute, 30(2):2631–2638 (**2005**).

[86] Williams, W. R., Zhao, J., and Schmidt, L. D. *Ignition and extinction of surface and homogeneous oxidation of NH3 and CH4*. AIChE J., 37(5):641–649 (**1991**).

[87] Scarpa, A., Barbato, P. S., Landi, G., Pirone, R., and Russo, G. *Combustion of methane-hydrogen mixtures on catalytic tablets*. Chemical Engineering Journal, 154(1-3):315–324 (**2009**).

[88] Pfefferle, L. and Pfefferle, W. *Catalysis in Combustion*. Catalysis Reviews - Science and Engineering, 29(2&3):219–267 (**1987**).

[89] Vlachos, D., Schmidt, L., and Aris, R. *Ignition and extinction of flames near surfaces: Combustion of H2 in air*. Combustion and Flame, 95(3):313–335 (**1993**).

[90] Vlachos, D. G., Schmidt, L. D., and Aris, R. *Ignition and extinction of flames near surfaces: Combustion of CH4 in air*. AIChE J., 40(6):1005–1017 (**1994**).

[91] Bui, P.-A., Vlachos, D., and Westmoreland, P. *Homogeneous ignition of hydrogen-air mixtures over platinum*. Symposium (International) on Combustion, 26(1):1763–1770 (**1996**).

[92] Springmann, S. *Kinetische Grundlagen und dynamische Simulation der autothermen Kraftstoffreformierung*. Ph.D. thesis, Insitut für Chemische Verfahrenstechnik, Universität Stuttgart (**2003**).

[93] Hickman, D. A. and Schmidt, L. D. *Synthesis gas formation by direct oxidation of methane over Pt monoliths*. Journal of Catalysis, 138(1):267–282 (**1992**).

[94] Tsang, S. C., Claridge, J. B., and Green, M. L. H. *Recent advances in the conversion of methane to synthesis gas*. Catalysis Today, 23(1):3–15 (**1995**).

[95] J. H. Lee, D. L. T. *Catalytic combustion of methane*. Fuel Processing Technology, 42:339–359 (**1995**).

[96] Klump, F. *Experimentelle Untersuchung der Zündvorgänge einer Brennkammer unter betriebsrelevanten Bedingungen*. Bachelorarbeit, Institut für Chemische Verfahrenstechnik, Universität Stuttgart (**2010**).

[97] Speidel, M. *Untersuchung der Strömungsführung in einer Brennkammer zum effizienten Wärmeeintrag in einen Festbettreaktor mittels CFD-Simulationen*. Diplomarbeit, Institut für Chemische Verfahrenstechnik, Universität Stuttgart (**2010**).

[98] *Ansys CFX*. `http://www.ansys.com/`.

[99] Lieb, S. *Experimentelle Untersuchung eines autothermen Festbettreaktors mit Strömungsumkehr zur Wasserstoffherstellung*. Diplomarbeit, Institut für Chemische Verfahrenstechnik, Universität Stuttgart (**2003**).

[100] Dieter, A. *Experimentelle Untersuchung eines ortsverteilten Systems für die temperaturkontrollierte Wärmeeinspeisung in Festbettreaktoren*. Diplomarbeit, Institut für Chemische Verfahrenstechnik, Universität Stuttgart (**2009**).

[101] *Gantner Instruments GmbH*. `http://www.gantner-instruments.com/`. Online; accessed 03-November-2010.

[102] Gose, T. *Kinetik der Methanreformierung: kinetische Messungen und Parameteranpassung*. Diplomarbeit, Institut für Chemische Verfahrenstechnik, Universität Stuttgart (**2005**).

[103] Eigenberger, G. *Chemische Reaktionstechnik II (Chemical Reaction Engineering II)*. Tech. rep., Lecture script, Institute for Chemical Process Engineering, University of Stuttgart (**2001**).

[104] Xu, J. and Froment, G. F. *Methane steam reforming, methanation and water-gas shift: I. Intrinsic kinetics*. AIChE J., 35(1):88–96 (**1989**).

[105] Kvamsdal, H., Svendsen, H., Olsvik, O., and Hertzberg, T. *Dynamic simulation and optimization of a catalytic steam reformer*. Chemical Engineering Science, 54(13-14):2697–2706 (**1999**).

[106] Pedernera, M. N., Piña, J., Borio, D. O., and Bucalá, V. *Use of a heterogeneous two-dimensional model to improve the primary steam reformer performance*. Chemical Engineering Journal, 94(1):29–40 (**2003**).

[107] Wesenberg, M. H. and Svendsen, H. F. *Mass and Heat Transfer Limitations in a Heterogeneous Model of a Gas-Heated Steam Reformer*. Industrial & Engineering Chemistry Research, 46(3):667–676 (**2007**).

[108] Poling, B. E. e. a. *The Properties of Gases and Liquids* (McGraw-Hill), 5th ed. (**2001**). Poling, B. E., Prausnitz, J. M. and O'Connell J. P.

[109] Warnatz, J. and Maas, U. *Technische Verbrennung* (Springer-Verlag) (**1993**).

[110] Krasnyk, M. *DIANA - An object-oriented tool for nonlinear analysis of chemical processes*. Ph.D. thesis, Max-Planck-Institut für Dynamik komplexer technischer Systeme (**2008**).

[111] *DIANA*. `http://www.mpi-magdeburg.mpg.de/projects/diana/`. Online; accessed 11-November-2010.

[112] Tränkle, F., Zeitz, M., Ginkel, M., and Gilles, E. D. *PROMOT: A Modeling Tool for Chemical Processes*. Mathematical and Computer Modelling of Dynamical Systems: Methods, Tools and Applications in Engineering and Related Sciences, 6(3):283–307 (**2000**).

[113] *ProMoT*. `http://www.mpi-magdeburg.mpg.de/projects/promot/`. Online; accessed 11-November-2010.

[114] Danner, T. *Anwendung der Modellierungs- und Simulationsumgebung Promot/Diana bei der rechnergestützten Untersuchung von Komponenten der Abgasnachbehandlung*. Studienarbeit, Institut für Chemische Verfahrenstechnik, Universität Stuttgart (**2008**).

[115] Danner, T. *Rechnergestützte Untersuchung von Systemen zur Abgasnachbehandlung mit ProMoT und DIANA*. Bachelorarbeit, Institut für Chemische Verfahrenstechnik, Universität Stuttgart (**2009**).

[116] Patzelt, P. *Beschreibung des experimentellen Betriebes eines autothermen Strömungsumkehrreaktors zur Methan-Dampfreformierung mittels Simulation*. Bachelorarbeit, Institut für Chemische Verfahrenstechnik, Universität Stuttgart (**2010**).

[117] Hindmarsh, A. C., Brown, P. N., Grant, K. E., Lee, S. L., Serban, R., Shumaker, D. E., and Woodward, C. S. *SUNDIALS: Suite of nonlinear and differential/algebraic equation solvers*. ACM Trans. Math. Softw., 31:363–396 (**2005**).

[118] Hindmarsh, A. C., Serban, R., and Collier, A. *User Documentation for IDA v2.6.0*. Tech. rep., Center for Applied Scientific Computing, Lawrence Livermore National Laboratory (**2009**).

[119] Lecheler, S. *Die Diskretisierung der Erhaltungsgleichungen*. In *Numerische Strömungsberechnung*, 41–60 (Vieweg+Teubner) (**2009**). ISBN 978-3-8348-9267-6.

[120] Frauhammer, J. *Numerische Lösung von eindimensionalen parabolischen Systemen mit adaptiven Gittern.* Diplomarbeit, Institut für Chemische Verfahrenstechnik, Universität Stuttgart (**1992**).

[121] *Cantera Project.* `http://code.google.com/p/cantera/`. Online; accessed 15-November-2010.

[122] Kolios, G. *Zur autothermen Führung der Styrolsynthese mit periodischem Wechsel der Strömungsrichtung.* Ph.D. thesis, Institut für Chemische Verfahrenstechnik, Universität Stuttgart (**1997**).

[123] Brevier. *Brevier Technische Keramik* (Verband der Keramischen Industrie e.V.) (**1998**).

[124] Burcat, A. and McBride, B. *1994 Ideal Gas Thermodynamic Data for Combustion and Air- Pollution Use.* Tech. rep., Technion Report TAE 697 (**1993**).

[125] Weiß, S. e. a. *Verfahrenstechnische Berechnungsmethoden - Teil 7 Stoffwerte* (Weinheim: VCH), 1st ed. (**1986**). Weiß, S., Berghof, W., Grahm, E., Gruhn, G., Güsewell, M., Plötner, W. Robel, H and Schubert, M.

[126] Froment, G. and Bischoff, K. *Chemical Reactor Analysis and Design* (John Wiley & Sons) (**1979**).

[127] Verlag, V., ed. *VDI Wärmeatlas, 7. Auflage* (VDI Verlag) (**1994**).

[128] Dixon, A. G. and Cresswell, D. L. *Theoretical prediction of effective heat transfer parameters in packed beds.* AIChE J., 25(4):663–676 (**1979**).

[129] Stegmaier, M. *Modular Simulation of Pressure Swing Adsorption for Hydrogen Purification in Compact Units.* Ph.D. thesis, Insitut für Chemische Verfahrenstechnik, Universität Stuttgart (**2008**).